PENGUIN BOOKS

WHAT AN OWL KNOWS

Jennifer Ackerman has been writing about science and nature for more than three decades. Her previous books include *The Bird Way*, the *New York Times* bestseller *The Genius of Birds, Birds by the Shore, Sex Sleep Eat Drink Dream*, and *Chance in the House of Fate*. Ackerman's articles and essays have appeared in *National Geographic, The New York Times Magazine, Scientific American*, and many other publications. She is the recipient of numerous awards and fellowships, including a National Endowment for the Arts Literature Fellowship in Nonfiction, a Bunting Fellowship, and a grant from the Alfred P. Sloan Foundation.

Praise for *What an Owl Knows*

"Ackerman is a warm and companionable guide, so enthusiastic about her subject that I suspect even the avian-indifferent will be charmed by her encounters with owls and the dedicated people who study them." —*The New York Times*

"A must read for all bird lovers, Ackerman's latest engaging work contains a feast of revelations about creatures that have fascinated us throughout human history. We learn that owlets begin making vocal sounds in the egg and that adult owls use sophisticated communication calls conveying their weight, sex, size, and state of mind. Ackerman describes the strong maternal instincts owls display and outlines the environmental threats to these entrancing birds."

—*The Guardian*

"[Ackerman] offers an absorbing ear-tuft-to-tail appreciation of the raptor that Mary Oliver, a poet, called a 'god of plunge and blood.' Owls, it seems, know a lot. Ms. Ackerman draws on recent research to explain what and how."

—*The Economist*

"Ackerman is an excellent writer on the natural world, and in this wise and thoroughly researched book, her chapter on human-owl relations is especially fascinating." —*New Statesman*

"Ackerman (our smartest bird writer) should do for hooting what a wave of octopus books did for slithering." —*Chicago Tribune*

"A fascinating read on how scientists are beginning to better understand the lives and ecology of these secretive and rarely visible birds." —*Science*

"A gripping history . . . *What an Owl Knows* is a treat." —*Financial Times*

"An enchanting guide." —*People*

"[*What an Owl Knows*] zips along merrily because Ackerman's love for these birds is totally infectious." —*Daily Mail*

"Lively and informative . . . While her straightforward style enables easy comprehension for the science-phobic, there is lyricism too. . . . Her sense of wonder runs through the book." —*The Times Literary Supplement*

"A rich and engaging installment of strigiphilia (i.e., love of owls). The owl-infatuated will find it delightful. The owl-curious will find it revealing. Ackerman is the author of previous books on avian intelligence behavior and on shorebirds, and she has a relaxed, assured, and informative voice. . . . Having seen more than a quarter of the world's nearly 250 species of owls myself, many chapters of Ackerman's book left me itching to hike snowy mountain trails in Montana, book airplane tickets to Serbia, and explore limestone cliffs in Australia."

—Richard Prum, *Yale Alumni Magazine*

"Few books can pull together academic summaries with the joy of seeing owls . . . Ackerman has been able to bring together the expertise of everyone interested in owls, from hobbyists, rehabilitators, veterinarians, photographers, owl-alcoholics, and researchers. She weaves together an enormous amount of information from an enormous number of people to give us perhaps the best popular and fact-filled informational book on owls to date."

—Denver Holt, founder, Owl Research Institute

"Owl fans, rejoice: Jennifer Ackerman's latest book, *What an Owl Knows: The New Science of the World's Most Enigmatic Birds,* is the ultimate guide to all things feathered and nocturnal. . . . Surprising and enthralling stories can be found on every page. . . . Those who enjoyed Ackerman's previous books on bird behavior and intelligence, *The Genius of Birds* and *The Bird Way,* will be pleased to find that Ackerman's writing is just as engaging and enthusiastic as ever; her genuine affection for her subjects infuses her descriptions of scientific studies of owls and the researchers who've carried them out. . . . *What an Owl Knows* is an ideal introduction to these enigmatic and fascinating birds, sure to convert a whole new generation of owl lovers." —American Birding Association

"Ackerman's account brings a sense of enchantment and wonder to this intriguing, complex animal, one that exerts an especially powerful pull on our psyche and culture." —*Australian Financial Review*

"'What is it about owls that so enthralls us?' . . . [Ackerman] explores this question with her trademark thoroughness and care, leading readers on an in-depth tour through the extraordinary world of owls. . . . Edifying and immersive."

—*Bookpage* (starred review)

"[A] masterful survey . . . There's fascinating trivia on every page . . . making for a revelatory glimpse into the lives of the 'enigmatic' raptors. Bird lovers will be enthralled." —*Publishers Weekly* (starred review)

"Fascinating food for thought for owl seekers and sure to please any lover of immersive treks into the lives of birds." —*Kirkus Reviews* (starred review)

"Always eloquent and engaging, science writer Ackerman turns her attention to owls, those mysterious, nocturnal birds that everyone can recognize but few really know. . . . Ackerman's latest vivid and compelling narrative is enlivened by her own passion for owls and her excitement over discoveries in the wild that show that, for humans, owls continue to be full of surprises. . . . Captivating."

—*Booklist* (starred review)

What an Owl Knows

The New Science *of the* World's Most Enigmatic Birds

Jennifer Ackerman

PENGUIN BOOKS

PENGUIN BOOKS
An imprint of Penguin Random House LLC
penguinrandomhouse.com

First published in the United States of America by Penguin Press,
an imprint of Penguin Random House LLC, 2023
Published in Penguin Books 2024

Illustration credits appear on pages 315–318.

ISBN 9780593298909 (paperback)

THE LIBRARY OF CONGRESS HAS CATALOGED THE HARDCOVER EDITION AS FOLLOWS:

Names: Ackerman, Jennifer, 1959– author.
Title: What an owl knows : the new science of the
world's most enigmatic birds / Jennifer Ackerman.
Description: New York : Penguin Press, 2023. |
Includes bibliographical references and index.
Identifiers: LCCN 2022053096 (print) | LCCN 2022053097 (ebook) |
ISBN 9780593298886 (hardcover) | ISBN 9780593298893 (ebook)
Subjects: LCSH: Owls—Popular works.
Classification: LCC QL696.S8 A25 2023 (print) |
LCC QL696.S8 (ebook) | DDC 598.9/7—dc23/eng/20230206
LC record available at https://lccn.loc.gov/2022053096
LC ebook record available at https://lccn.loc.gov/2022053097

Printed in the United States of America
1st Printing

Designed by Amanda Dewey

To my sister Nancy, with love and gratitude

CONTENTS

Preface: Who Knew *xi*

One.

MAKING SENSE OF OWLS:

Unpacking the Mysteries *1*

Two.

WHAT IT'S LIKE TO BE AN OWL:

Ingenious Adaptations *13*

Three.

OWLING:

Studying the World's Most Enigmatic Birds *47*

Four.

WHO GIVES A HOOT:

Owl Talk *79*

Five.

WHAT IT TAKES TO MAKE AN OWL:

Courting and Breeding *113*

Six.
TO STAY OR TO GO?:
Roosting and Migrating *163*

Seven.
AN OWL IN THE HAND:
Learning from Captive Birds *203*

Eight.
HALF BIRD, HALF SPIRIT:
Owls and the Human Imagination *233*

Nine.
WHAT AN OWL KNOWS:
How Wise Are Owls? *255*

Afterword.
SAVING OWLS:
Protecting What We Love *273*

Acknowledgments 295
Further Reading 301
Illustration Credits 315
Index 319

PREFACE

Who Knew

Long-eared Owl

What is it about owls that so enthralls us? They appear in the Chauvet Cave paintings of France dating to 30,000 years ago and in the hieroglyphics of ancient Egyptians, in Greek mythology and among the deities of the Ainu people of Japan, in the prints and etchings of Picasso and as couriers in the Harry Potter stories, shuttling between the realm of matter-of-fact Muggles and the magical. They inhabit our languages and are embedded in our sayings. When we're cranky, stubborn, uncooperative, we are "owly." If we stay up late or are active at night, we're "night owls." If we're aged and sage, we're "wise old owls."

In some places, owls vie with penguins for popularity. In others, they're vilified as demon spirits. Owls have this kind of duality. They're tender and deadly, cute and brutal, ferocious and funny, sometimes even playing the mischievous clown, stealing camera equipment or snatching hats. We see something deeply familiar in them, with their round heads and big eyes, and at the same time, an intimation of a whole other kind of existence, the dark side of the one we inhabit. Most owls are nocturnal creatures that move about unseen, revealed only by their weird night hoots and cries. Their flight is velvety quiet, and their hunting skills, often deployed in pitch black, inspire awe.

In many cultures, owls are deemed half bird, half spirit, crossovers between the real and the ethereal, considered by turns symbols of knowledge and wisdom on the one hand, and bearers of bad luck and illness, even death, on the other. They're often viewed as prophets or messengers. The Greeks believed that an owl flying over a battlefield predicted victory. In the early folklore of India, owls crop up as symbols of wisdom and prophecy. So, too, among the Navajo. The Navajo myth of Nayenezgani, the creator, reminds people that they must listen to the voice of the prophet owl if they want to know their future. The Aztecs considered owls a symbol of the underworld, and the Maya, as messengers of Xibalba, the "place of fright." In *Julius Caesar*, Casca is terrified when an owl appears by day as an omen of imminent death: "The bird of night did sit, / Even at noonday upon the market-place, / Hooting and shrieking."

Owls exist on every continent except Antarctica and in every form in the human imagination. Yet for all this ubiquity and interest, scientists have only lately begun to puzzle out the birds in deep detail. Owls are much more difficult to find and study than other birds. They are cryptic and camouflaged, secretive and active at a time when access to field sites is challenging. But lately researchers have harnessed an array of powerful strategies and tools to study them and unpack their mysteries.

This book explores what new science has discovered about these enigmatic birds—their remarkable anatomy, biology, and behavior and the hunting skills, stealth, and sensory prowess that distinguish them from nearly all other birds. It looks at how researchers have pulled back the curtain on how owls communicate, court, and mate, how they raise their young, whether they act more from instinct or from learning, why they move from place to place or stay put to weather the seasons, and what they have to tell us about their nature—and our own. It explores new insights gleaned from studying owls in the wild and also in captivity, birds kept "in the hand," most often because they've been injured. Specialists who live and work with owls in intimate partner relationships are learning things one can learn only up close, one on one with an owl. They're advancing the science of caring for these birds, and in return, the owls they heal are helping to educate the public and revealing some of the deepest mysteries about their communication, their individuality and personality, their emotions and intelligence.

In analyzing the apparently "simple" hoots and calls of owls, for instance, researchers have found that their vocalizations follow complex rules that allow the birds to express not only their needs and desires but highly specific information about their individual identity, and their sex, size, weight, and state of mind. Some owls sing duets. Others duel with their voices. Owls can recognize one another by voice alone. Their faces are expressive, too. They may seem to wear the same bland meditative visage, imperturbable as the moon, but their appearance can change along with their feelings—a fascinating window on their minds, if you know how to read it.

Some owls migrate but not like other birds, and not in predictable patterns. Some owls cache or hoard their prey in special larders. Some decorate their nests. Burrowing Owls* live in underground burrows, sometimes alongside prairie dogs, and when threatened, will hiss like a cornered rattlesnake. They festoon their nests with corncobs, bison dung, shreds of fabric, even pieces of potato. Long-eared Owls roost in huge colonies, which, like the colonies of Cliff Swallows, may act as information centers. Scientists studying barn owls have discovered that the baby owls sleep like baby humans, spending more time in REM (or dream) sleep than adult owls. Why? Can owls help us determine the role of REM sleep in brain development in both birds and humans? Do owls talk in their sleep?

Most owls are socially monogamous, pairing up to breed, but research suggests they're also genetically monogamous—unlikely to engage in extra-pair copulations—highly unusual in the bird world. This may be so, but are they as loyal to their mates as we imagine?

Owls are known as "wolves of the sky" for more good reasons than ever. Fierce hunters, they take all kinds of prey, from mice and birds to opossums and small deer, and even other owls. But they also occasionally scavenge, everything from porcupines to crocodiles and Bowhead Whales. Elf Owls dine on scorpions—only after they remove the venomous stingers—and, like other owls, get most of the water they need from their prey. Stygian Owls, which prey primarily on birds, have figured out how to find a whole night's feast in a single swoop. According to Brazilian ornithologist José Carlos Motta-Junior, the owls use the noises of gregarious group-roosting birds like Blue-Black Grassquits to zero in on them and then, one by one, take the whole assembly. "I found pellets with the remains of five or more grassquits, my record being a pellet with eleven!"

---

* In general, common names of species are capitalized in this book (e.g., Long-eared Owl, Eastern Screech Owl, Osprey). However, if there is more than one species in a genus, the term that applies to the group as a whole is not capitalized (e.g., barn owls, screech owls, kingfishers).

Groundbreaking work on owl senses is shedding light on the superpowers that allow these birds to find their prey at night—the strange features of their superb night vision and hearing, their extraordinary ability to locate noises, their near-soundless flight—adaptations that make owls a pinnacle not just of the food chain but of evolution itself. Owls may have lost some ability to distinguish color over evolutionary time, but they have exquisite sensitivity to light and movement. They can see ultraviolet light, too, with equipment that differs dramatically from most other birds. Better understanding of owl ears, described as the "Ferraris of sound sensitivity," has shifted our view of their superhuman hearing and even shaped hearing tests for babies. Scientists have parsed the unexpected ways a Great Gray Owl performs a stunning feat in winter—catching voles hidden deep beneath snow by sound alone. A new view of the way owls *process* sound has also yielded news: some of the sounds owls perceive are processed in the visual center of their brains, so they may actually get an optical picture of a sound—a mouse's rustle flashing like a beacon in the forest dark. And here's a discovery to boggle the mind: an owl's brain uses math to pinpoint its prey. Who knew.

To my mind, these findings don't diminish the wonder of owls, they intensify it.

An owl is an owl is an owl.

Not so. Owls vary dramatically from species to species and even from individual to individual within a species. It's one of the reasons I wanted to write about this order of birds—to explore the idiosyncrasies of different kinds of owls and what has been discovered about their evolution, species adaptations, and individual natures.

So many of the generalizations we've learned about owls don't hold true for all species. Not all owls are nocturnal. Not all fly quietly. Not all have asymmetrical ears. Not all mate for life. Not all roost in forest trees. Some owls roost in caves, like the Australian Masked Owl of the Nullarbor Plain, others on the ground or beneath it, like Burrowing Owls. I saw my first Powerful Owl—a rapacious hunter and the only Australian bird able to carry more than its body weight in prey—perched in an urban tree in the middle of Sydney. Some owls, like the Spectacled Owl, summon visions of deep rainforests. Others, like the Snowy Owl or the Boreal Owl, conjure icy northern landscapes. Why are Snowy Owls white? It's not as simple as it seems.

Owls are not only cryptic, guarded, and secretive, they're also dissidents and iconoclasts, rule breakers. We think of owls as solitary, for instance, but a few species congregate—like those Long-eared Owls that roost in big colonies. In tropical regions, owls may form communities with up to seven different species living together. The Mottled Owl of Central and South America has been known to hold meetings of several individuals during the night—a true parliament of owls—for a still mysterious purpose.

Owls may be known for their nocturnal way of life, but only about a third of owl species hunt solely at night. Others hunt at dusk. Great Gray Owls are mainly nocturnal but hunt in the daytime during the breeding season, when they must feed their young. Other species, like Northern Hawk-Owls and Northern Pygmy Owls, hunt in the day all year round. If you're lucky, you might see a Northern Hawk-Owl hunting by sight in the openings of boreal forests in the far north, spotting its prey up to a half mile away, then swooping down from a tree perch or even hovering like a kestrel to snatch up a small bird or shrew.

Northern Pygmy Owls are renegades in another way as well. Most owls lay their eggs over the course of several days, and their chicks hatch at different times. Northern Pygmy Owls, it seems, buck convention, hatching their chicks all at once.

Owls set my head a-whirr with questions. Why do they wield such a hold on the human imagination? They have a reputation for wisdom, but are they smart? Do they act by instinct alone, or are they curious and inventive? Do they have feelings and emotions? Why do an owl's eyes, alone in the bird world, face the same way ours do? What made the early ancestors of owls cross the boundary into night? And why do some owls hunt during the day? Owls live all around the globe, but there are hot spots of owl diversity—in southeastern Arizona and western Mexico, southern Asia, southeastern Brazil. What draws so many species to these places? How are owls adapting to shifts in their habitat and global climate?

Throughout this book you'll find discoveries that answer these questions and pose others. You'll find the insights and observations of vets and educators familiar with the intimate lives and habits of owls, ethnoecologists exploring the deep hold of these birds on our psyche, and biologists and ecologists investigating their importance in the natural world and how we can best preserve them. You'll also find portraits of people obsessed with owls, some famous—such as Florence Nightingale, Teddy Roosevelt, Pablo Picasso—and some not, like the librarian at the Metropolitan Museum of Art who collects owl images from throughout history and wears a particularly beautiful one on her body. You'll encounter citizen scientists who have boosted owl research, "ordinary" people not trained as researchers but contributing in brilliant ways to our knowledge of owls. One Dutch musician uses her finely tuned musical ear to listen for individuality, infidelity, and divorce among Eurasian Eagle Owls. A heart surgeon turns his intense focus to the intimate conversation between pairs of Northern Pygmy Owls, what he calls "soft talk," to understand their courtship and pairing. An emergency room nurse bands migrating Northern Saw-whet Owls through the night, providing balm for the trauma of her job and hard data on the movements of these elusive little owls—once thought rare, now recognized as surprisingly common in large part because of volunteers like her.

And, of course, you'll meet the scientists and researchers who have devoted their lives to understanding these birds. When I asked David Johnson, who has studied owls for well over forty years and directs the Global Owl Project, why he loves owls, he told me, "I didn't choose them. They chose me." Good thing they did. Johnson and his team of 450-plus researchers from around the world have been working together for the past two decades to conserve all the planet's owl species.

But the real heroes of this book are the owls themselves. For millennia we have looked to these birds as messengers and signs. What are they telling us now?

"If anyone knows anything about anything," says Winnie-the-Pooh, "it's Owl who knows something about something." Owls have truths to tell us, from afar—from their perches and nests deep in old-growth forests, deserts, the Arctic—and from up close, in the hands of vets, rehabbers, researchers, and educators. We would be wise to listen.

One

Making Sense of Owls

Unpacking the Mysteries

Owls are probably the most distinctive order of birds in the world, with their upright bodies, big round heads, and enormous front-facing eyes—hard to mistake for any other creature. Even a young child has little trouble identifying them. The same is true for a range of species, including other birds—chickadees and titmice, ravens and crows—which can spot the shape of an owl instantly and single it out as an enemy. But beyond the basics of that telltale form, what makes an owl an owl? And how did these extraordinary birds get to be the way they are?

Through research on owls past and present, scientists are tracing these birds back to their earliest beginnings to make sense of their evolution and their family tree. Owls first appeared on earth during the Paleocene epoch, some fifty-five million to sixty-five million years ago. Tens of millions of years later, they split into two families, Tytonidae (barn owls) and Strigidae (all other owls). Like all birds, they initially arose from a group of small, mostly predatory, running dinosaurs that were coexisting with other, larger dinosaurs sixty-six million years ago. That all changed when an enormous asteroid struck earth, triggering the mass extinction that killed off most of the big land-based dinosaurs. A few of the bird ancestors survived, including the forerunners of today's owls and all other living bird species.

As a group, owls were initially thought to be related to falcons and hawks because they shared a hunting lifestyle like these raptors. Later, they were lumped with nocturnal birds such as nightjars on account of their big eyes and camouflaged plumage. But new research shows that owls are most closely related not to falcons or nightjars but to a group of day-active birds that includes toucans, trogons, hoopoes, hornbills, woodpeckers, kingfishers, and bee-eaters. Owls probably diverged from this sister group during the Paleocene, after most of the dinosaurs died off and small mammals diversified. Some of those little mammals took to night niches, and owls adapted, evolving a suite of traits to take advantage of the nocturnal feast. Now most owls share an array of remarkable features that distinguish them from other birds and give them a unique ability to hunt at night, including retinas rich in cells that provide good vision in dim light, superior hearing, and soft, camouflaged feathers tailored for quiet flight. Of the 11,000 or so species of birds alive today, only 3 percent have these sorts of adaptations that allow for stalking prey in the dark.

Since their first appearance on the planet, about a hundred owl species have come and gone, leaving fossil traces of their existence, including *Primoptynx,* a peculiar owl that soared across Wyoming skies fifty-five million years ago and hunted more like a hawk than an owl, and the Andros Island Barn Owl, a full three feet tall, which terrorized Pleistocene mammals. One extinct owl that vanished from the Indian Ocean island of Rodrigues relatively recently, in the eighteenth century, had a smaller brain than most present-day owls but a well-developed olfactory sense, suggesting it may have used its nose more for hunting and perhaps even scavenging.

Some 260 species of owls exist today, and that number is growing. They live in every kind of habitat on almost every continent—from desert and grassland to tropical forest, mountain slopes, the snowy tundra of the Arctic—and they range widely in size, appearance, and behavior, from the diminutive Elf Owl, a little nugget of a bird, impish, troll-like, about the size of a small pine cone and the weight of eight stacked nickels, to the massive Eurasian Eagle Owl, which can take a young deer; from the delicate Northern Saw-whet Owl that "flies like a big soft moth," as Mary Oliver wrote, to the comical, slim-legged Burrowing Owl, with its bobbing salute. There are Chocolate Boobooks and Bare-legged Owls, Powerful Owls and Fearful Owls (named for their bloodcurdling, humanlike scream repeated every ten seconds), White-chinned Owls and Tawny-browed Owls, Vermiculated Screech Owls and Verreaux's Eagle Owls, Africa's biggest, with its startling pink eyelids. Some owls, like the ubiquitous barn owls that occur in multiple forms worldwide, carry a raft of common names reflective of their mythic power: demon owl, ghost owl, death owl, night owl, church owl, cave owl, stone owl, hobgoblin owl, dobby owl, monkey-faced owl, silver owl, and golden owl.

Much to the amazement of researchers, new owl species are still turning up, including an owl that stunned scientists when it was discovered high in the Andean mountains of northern Peru. The Long-whiskered Owlet, a tiny, bizarre owl—one of the rarest birds in the world—with long wispy facial whiskers and stubby wings, is so different from other owls that scientists put it in its own genus, *Xenoglaux*, which means "strange owl" in Greek. It sings a rapid song described as "low, gruff, muffled *whOOo* or *hurr* notes" and is found only in high forests between two rivers in the Andes. In 2022, scientists discovered a new species of scops owl on the island of Príncipe, off the west coast of Africa, named *Otus bikegila* for the park ranger who was instrumental in bringing it to light. Because some owls live in isolated regions like

Long-whiskered Owlet

these, in tropical rainforests and on mountains and islands where populations separated geographically can diverge genetically, the number of species may continue to climb.

Also boosting the species count and shifting the owl family tree is a deeper understanding of already recognized owl species. By closely examining the body structures, vocalizations, and DNA of known owl species, scientists are finding sufficient differences between populations to split one species into two or more.

Take barn owls. The oldest lineage of owls, barn owls probably first arose in Australia or Africa, spread through the Old World, and now live on nearly every continent. Because they look alike over their entire range, they were once classified as a single species. But owls are showing us that appearances can be deceiving. DNA studies have revealed that Tytonidae, the scientific name for barn owls, is in fact a rich complex of at least three species, with a total of some twenty-nine subspecies. And there may be others existing in remote places that haven't yet been recognized. Likewise, researchers recently used genetics to tease apart two new screech owl species from Brazil that had been grouped with other South American species: the Alagoas Screech Owl of the Atlantic rainforest, and the Xingu Screech Owl found in the Amazon. Both owls are threatened by deforestation and are at risk of extinction.

Along with new species, a flock of new insights on the nature of owls has flowed from laboratories and field studies around the world in the past decade or so, shedding light on a profusion of owl mysteries. Why are these discoveries emerging now? How are scientists making sense of the hidden lives and habits of these inscrutable birds?

For one thing, there are innovative new tools for studying the evolution, anatomy, and biology of owls and for finding them in the wild, tracking their movements, and monitoring their behavior. Cutting-edge imaging technology such as X-ray computed tomography (CT) scanning allows researchers to see inside the bodies of living owls, visualizing the anatomical structures that relate directly to behavior, and to peer through rock to see into fossils. DNA analysis is revealing relationships in the owl tree of life, challenging old concepts about who is related to whom and how closely. New "eyes" in the field—infrared cameras and other night vision equipment, radio tagging, and drones over areas as remote as the snowy landscapes of Siberia—are advancing new discoveries about owl behavior or confirming older observations by banders and biologists who have been in the field for decades. Satellite telemetry is illuminating the movements of owls over short and long distances. Tiny satellite transmitters packed onto the backs of Snowy Owls, for instance, are revealing wondrous new insights on their mysterious movements, such as the puzzling northward journeys of some of these iconic birds in the dead of winter.

Nest cams are offering a look at intimate owl interactions at the nest that would otherwise be impossible to observe: the feeding of mates and young, for instance, and the squabbling between siblings.

"Nest cams tell all," says ornithologist Rob Bierregaard, who studies Barred Owls. "They offer the best picture of what's for dinner—flying squirrels, cardinals, salamanders, fish, crayfish, big insects—and how the feeding goes. You can see the male handing over food to the female to feed the young. I've seen males stash mice on branches and possum, too, delivering it piece by piece." Nest cams expose the sometimes nasty, sometimes charitable dynamics between siblings. Chicks in a brood can be selfish and competitive, to the point of siblicide. But some owlets display a remarkable form of altruism rare in the animal world. Nestling barn owls, for instance, are known to give food over to their younger siblings, on average twice per night.

Biologist Dave Oleyar pursued his master's research in the late 1990s and says he wishes he had had today's technology back then. "It's amazing what we can do now," he says. "Running these nest cams 24-7 and documenting prey deliveries to the nest, what the parents are bringing in and how often, we can gather a huge amount of data about their foraging patterns. Before we had these 'eyes' in the field, the logistical challenges of studying nestling growth, development, and interactions were overwhelming and limiting."

Listening to owls remotely with sophisticated new audio recording devices has been a boon to owl research, helping scientists understand the interplay of different owl species without disturbing them. With acoustic monitoring, for instance, researchers are sorting out the dynamics between Barred Owls and endangered California Spotted Owls of the Sierra Nevada. In placing audio recorders in close to a thousand locations across 2,300 square miles of mountainous terrain to collect owl calls, they have discovered completely unexpected interactions between the aggressive Barred Owls, on the one hand, and the smaller, but still surprisingly feisty spotted owls, on the other—with significant implications for conservation.

Another unusual new method for surveying and monitoring owls is distinctly less high tech and more nose heavy. Researchers are harnessing the olfactory powers of dogs to locate elusive owl species in places as far-flung as Tasmania and the Pacific Northwest. Specially trained "sniffer" dogs snuffle the pellets, those misshapen cigars made of leftover bits of undigested fur and bone, which owls eject onto the ground beneath their roosts and nests. The pellets are hard to spot, but they emit odors so the dogs can easily sniff them out, leading a researcher straight to the spots where the owls hang out.

Many breakthroughs have also come from more traditional ways of studying owls—trapping, measuring, and banding them—and monitoring the birds over long periods of time. Long-term study of owls in the wild is slow, hard work in all weathers, season after season, year after year, but it's yielding vital new windows on breeding behavior and population trends. Decades-long studies of Long-eared Owls, Burrowing Owls, Snowy Owls, and Tawny Owls are revealing how owls are responding to habitat loss and climate change, pointing us toward avenues for conservation, not just of owls but of whole ecosystems.

Making sense of owls means witnessing them in the wild, in their natural habitat. But while owls may be easy to recognize, they're not easy to see, even for the experts. They often hide right under our noses in the day, camouflaged against the bark of trees or tucked into hollows, and in the night, sail off into the darkness, unobserved. "Finding owls is hard," says David Lindo, a naturalist, photographer, and highly experienced bird guide known as the Urban Birder, who is forever on the lookout for birds. "It's often a question of diligence. You have to commit yourself to it. You have to try and work out where they are and then religiously search the trees, look for pellets and splashback [the feces of owls, also known as whitewash]."

This is why the sophisticated new tools for owl detection and

Camouflaged Eastern Screech Owl

monitoring are so vital. But even with these powerful technologies, locating owls in the wild is still often a maddening and elusive treasure hunt. As Sergio Cordoba Cordoba, an ornithologist studying neotropical owls, told me, "It can be really frustrating. Technology is a great ally, infrared cameras and telemetry, but we often still rely on sounds. Trying to find an owl you hear singing is like being an explorer of old times. You try to follow the sound, walk or crawl to get nearer without making any noise (almost impossible with dry leaves on the forest floor), and when you think you are near enough, switch on the flashlight and see who is singing. Most times, I flush the owner and never find out who it is!"

Researchers and birdwatchers often attract owls with "playback," using audio recordings of owl territorial or mating calls to draw them in. "A guide may play the call of a particular species, like a screech owl," as Lindo explains, "and then five minutes later, one pops up in the tree, you flash a torch, take a picture, and then it's gone." I had the thrill of seeing a family of Striped Owls and two species of neotropical screech owls in southeastern Brazil using this method. It's an important tool for researchers. But as Lindo says, for the casual birdwatcher, "it's a bit of a cheat" and can disrupt the owls' natural behaviors.

Nothing beats a chance encounter, happening across an owl in the wild. People who understand the privilege of stillness and just sit, look, and listen—like owls themselves—sometimes get lucky. One of Lindo's most memorable owl moments came about this way. Some years ago, he was leading a bird tour in Helsinki, Finland. He had a day to himself, so he borrowed a bike from the hotel. "I noticed that there was a green area of woods near to me on an island," he told me. "So I cycled across a bridge to the island. I remember putting my bike down and just sitting in the forest. As I sat there, a Great Tit came really close to me. It landed on my cap, and then darted back up to the tree. It did this a couple of times, which puzzled me. Then I noticed something swoop across the clearing in front of me. It was a young Long-eared Owl, and it was hunting, totally unaware of me. I just sat there and watched it for maybe forty minutes, flying around, sometimes stopping very close to me. I kept stock-still. I was camouflaged by the trees, and it didn't notice me at all. That was an amazing moment."

Jennifer Hartman, who spent years studying Northern Spotted Owls, describes sitting silently and observing the endangered birds one on one for up to eighteen hours at a time. "I didn't think a person could spend time with a wild owl that way and not have them be stressed out or fly away," she says. "Sometimes they would fall asleep while I was there. I saw owls drop to the forest floor and sip water from a pool. I saw them wake up from a nap and flutter down to the forest floor and spread their wings out in a patch of sunlight—maybe to drop mites from their feathers or let ants up onto their feathers to eat the mites. Once I watched a hummingbird dive-bomb a female owl who was sleeping. And she woke up and was like, 'What the heck? I'm not doing anything!'

"And the sounds they made were extraordinary, too," she recalls. "When a goshawk flew by, the male would make this low call that I hadn't heard ever before, and it was his alert to the female, 'Stay low, hunker down, don't move.' I was learning all these different things about them that I couldn't have learned in a book. It was just this very intimate, very quiet, almost otherworldly experience, and it changed my life."

Owls change lives, and the effort to make sense of them shapes how we experience the world, heightens our wonder.

I saw this clearly one spring day in a gully dense with hawthorn and Chokecherry trees in western Montana. I held a female Long-eared Owl caught in the wild, my palm curled around her feet, her talons tucked between my fingers. Owl expert Denver Holt was at my side, guiding her release. "Watch closely when she goes," he whispered. It had taken us all morning and a good part of the afternoon to trap her in the mist nets. She was a big bird, mature, wary, not easily captured, with strong legs and feathers as soft as rabbit fur.

Earlier I had spotted her through my binoculars when she was roosting, lying low as owls often do in the day. At first, I couldn't grasp what I was looking at, a thin, dark mass in the tangled branches of a hawthorn tree that seemed to vanish every time I looked away and then looked back. There and not there. I thought my eyes were playing tricks on me. She looked less like an owl and more like a broken branch, utterly still, upright, and rigid, stretched vertically with her girth so contracted she appeared perfectly cylindrical, very thin and very tall. Her long ear tufts that give her kind its name were fully extended, tight and parallel, a way of disrupting that telltale owl contour so she blended seamlessly with the branches of her roost. She was a warm grayish brown with a mottled mix of horizontal and vertical streaks just like tree bark. If it hadn't been for her eyes, a flaming yellow, I wouldn't have believed my own.

We had had to work hard to trap her, moving in from several different angles multiple times to flush her into the nets. When we finally did, and I held her, she locked eyes with me in a catlike stare. Now she was measured and weighed, banded, and ready for release. I crouched low in the tunnel of alders and pointed her toward a narrow opening through the thick snarly branches, cocked my wrist a little, and then opened my fingers. She lifted off without a sound, spread her wings, and with slow, even beats, navigated through the narrow opening without so much as a whoosh, and vanished again into the dark thicket.

Holt has experienced thousands of such moments with owls. For me, it was an adventure, bright, intense, deeply affecting. That owl seemed like a messenger from another time and place, like starlight. Being near her somehow made me feel smaller in my body and bigger in my soul.

I asked Holt why he has devoted the better part of his life to studying these elusive creatures. Because of this, he said, gesturing at her empty path. Because they're so beautifully adapted to their world, so quiet, invisible, cryptic not just in coloring but in sound, deft in the dark, superb hunters—traits that have evolved over millions of years. "And," he said, "because they're still so full of surprises."

Two

What It's Like to Be an Owl

INGENIOUS ADAPTATIONS

WOLVES OF THE SKY

Soft twilight fading to dark in the bushland south of Sydney, Australia. Spring settles into a quiet night sky. Along the limbs of a fig tree, a Brushtail Possum trundles along, stops, gets busy with a ripe fig clutched hard between its little handlike paws. The possum is a big marsupial, about the size of a cat, with a pointy snout, pin nose, big ears, and a bushy black tail. The fruit is ripe, delicious, and worth pausing to finish off in the near dark. Suddenly, seemingly out of nowhere, a flurry of wings, then piercing talons, a steely grip. A loud squeal, a flailing struggle, a fatal bite in the neck. From the possum's perspective, it's over. The creature is consumed on the spot, headfirst.

This is the tidy work of a Powerful Owl, Australia's largest owl and a masterful predator. That word *predator* "is baggy with misuse," writes author J. A. Baker. "All birds eat living flesh at some time in their lives. Consider the cold-eyed thrush, that springy carnivore of lawns, worm stabber, basher to death of snails." Baker is right, of course. But owls are something else, pure hunters, ruthless, often

Powerful Owl with possum

gruesome in their eating habits. The Powerful Owl I saw was perched at the top of a eucalyptus tree in the botanic gardens at the heart of Sydney. Beneath the tree was a creamy smattering of whitewash and a large gray finger of pellet stuffed with fur and bone, likely all that remained of a possum or fruit bat. An owl like this will eat an astounding 250 to 350 possums a year, nearly one a day.

How it handles its food is a marvel itself. Possums it will often quickly disembowel in less than twenty seconds before downing the rest of the meal in large pieces. Possums are plant eaters; the owl has no use for all that vegetation and may not be able to digest it. Smaller prey it will swallow whole. As with all owls, the indigestible parts—fur, bones, feathers, claws—are sequestered in the stomach and compressed into a pellet. The pellet remains there for hours until the owl regurgitates it, pushing it back up the esophagus and into the mouth for ejection.

Burrowing Owl ejecting a pellet

This remarkable ability to move indigestible food up and out, against the usual direction, is called "antiperistalsis." Pterosaurs, those flying predators of the dinosaur era, could do it, too. The effort can be quite strenuous, which is why owls sometimes look as if they're wincing when they cough up a pellet. But it's an essential part of the digestive process: because the pellet blocks part of the digestive tract, an owl usually can't eat again until it's expelled.

Owls hunt and eat all kinds of animals. Some species are specialists, such as fish owls, which are almost exclusively piscivores, and Flammulated Owls, which eat mainly insects. Some, like Short-eared Owls and barn owls, favor voles and other small-rodent prey. But many owls are generalists and will hunt everything from spiders, frogs, salamanders, and mice to birds and occasionally bats. Some species, like the Ferruginous Pygmy Owl, are lightning predators, so swift and agile they can snatch a hummingbird on the wing while it's probing a flower. The Northern Hawk-Owl perches and pounces. Short-eared Owls course back and forth low over an open field or grassland, systematically quartering the ground to detect voles, mice, and other small mammals. Undaunted by big prey, the Great Horned Owl is known to take woodchucks, rabbits, even domestic cats, and won't turn its nose up at skunks. As for birds, it will pluck up ducks from the water at night, and no goose is too big for it to tackle. Other owls are fair game, too—Long-eared and Barred and all the little forest owls—which makes the Great Horned Owl a superpredator, a predator that eats other predators.

Even Snowy Owls, which are renowned for their focus on the little Arctic rodents known as lemmings, turn out to be catholic in their appetites. You can tell a lot about what a raptor eats from its feet. "You look at the feet of a true small-mammal specialist, like a Rough-legged Hawk, these tiny little delicate feet," says ornithologist Scott Weidensaul. "And then you look at a Snowy Owl, which has these massively powerful feet. That is not a lemming specialist. This is an 'anything-I-can-shove-down-my-throat' specialist," including a sizable duck like an Eider or even a decomposing bottlenose dolphin.

For a long time, scientists thought owls didn't scavenge—and if they did, it was a fluke. But lately camera traps have caught owls helping themselves vulturelike to carrion of all kinds—Eurasian Eagle Owls feeding on deer and sheep, a Great Gray Owl feasting on an ungulate killed by wolves, a Long-eared Owl in Italy helping himself to four dead Crested Porcupines, a Snowy Owl gorging on a Bowhead Whale carcass in the Arctic, and a Brown Fish Owl dining on a crocodile corpse.

But owls take the lion's share of their prey live, and it's not easy. Most predators fail to capture their prey more often than they succeed. Possums and lemmings and voles don't just hang there on a vine waiting to be eaten. They hide or try to get away or even fight back. A

Brushtail Possum may stand on its hind limbs, forepaws close to its chest, growl, and then charge. Sometimes birds fight back en masse, mobbing and harassing owls until they give up their perch.

One testament to the hunting prowess of owls is their caching of surplus prey. Owls routinely cache, or hide, excess food in a nest, tree hole, or forked branch as a way of holding on to a glut of prey for later consumption. Caching most often occurs when female or young are satiated, and the male hides the leftovers. Owls sometimes kill more than they can initially eat when prey becomes easily available—like a sleeping cluster of Blue-Black Grassquits. Little Owls will return to a songbird nest again and again until all the young are taken. Northern Saw-whet Owls sometimes behead their prey, mostly mice and small birds, caching the body for future consumption. Pygmy owls in Norway have been known to stockpile as many as a hundred items (mostly small mammals) in a single larder to get through harsh winters.

In some ways, owls hunt like other raptors, pursuing their prey with strong talons and sharp beak. They have powerful muscles in their legs and feet and big talons, the better to seize and kill prey. We think of owls as short legged because they tuck in their legs at rest and in flight. But most have long, well-muscled legs, up to half the length of their bodies, with strong bones, especially in their feet. Just before contact with their quarry, they thrust their powerful feet forward to strike, killing their prey with the force of the impact and crushing talons. One recent study showed that an owl weighing less than a pound pouncing on a mouse can exert force equivalent to 150 times the weight of its prey. Sometimes, with bigger prey, an owl will kill by biting the neck with that sharp beak, or exerting a prolonged, suffocating squeeze with its feet. Owls lock on to their target with maximum power using two ingenious foot adaptations. They have four toes, three of which face forward in flight and sometimes in perching. But when owls need to grasp their prey, a special flexible joint allows them to swivel one rear toe forward to give them an extra powerful X grip. They can hold that grip without tiring thanks to a system of tendons in their feet that keeps the toes locked around prey without the exertion of muscles so they don't have to put energy into holding it. This also benefits owls that capture prey "blindly," beneath snow or leaves or in the complete dark, allowing them to lock on tight to their target even if they can't see or judge its exact size or shape.

Hunting is challenging for any raptor. But hunting at night? It's this power to find and take prey in the dark that's so unique to owls.

ALL EARS

I once had the chance to meet a Great Gray Owl up close. Percy was the male in a resident pair of Great Grays at Skansen, an open-air museum in Stockholm, Sweden. The zookeeper let me into the spacious aviary planted with trees and boulders and told me to stand quietly by a railing. At first the big bird stayed in a far corner of the enclosure. I could barely make him out against the tree bark, and even in this enclosed space, his partner was invisible. But when the zookeeper brought out a bowl of frozen mice, Percy launched, and with slow, quiet wingbeats, flew over to the railing and landed soundlessly just two or three feet from me. He looked enormous, and his massive head turned toward me until the whole round dish of his humanlike face hit me square. He was so close I could see his pupils, the dark holes in the centers of his eyes—marigold in the dark gray of his facial disk. When the zookeeper dipped his hand in the bowl, those eyes seemed to widen, and his head swiveled back to the bowl. The zookeeper gave him a frozen mouse, which he wolfed down. Then another, and another, swallowing each of them whole.

Great Gray Owls don't usually get their meals dished up like this in broad daylight. They rely on their skills as canny night hunters. A few other kinds of birds, like nightjars, potoos, and frogmouths, hunt large flying insects in night skies. But no other bird hunts mammals and other birds at night the way owls do.

Some years ago, two field naturalists in Canada watching Great Gray Owls hunting on dark winter nights noted how they would move from perch to perch until they reached a spot where they seemed to sense something beneath the snow.

"The bird would cease all scanning and would peer down at a sharp angle," the naturalists wrote. "It appeared almost hypnotized by the spot below and was very difficult to distract. . . . Although we were often observing from 10 to 20 feet away, very rarely did we see anything . . . yet the owls would almost invariably capture prey upon plunging into what appeared as bare snow."

Pinpointing invisible prey? What sort of magical power is that?

Roger Payne was the first to show that barn owls can hunt their prey in complete darkness using only sound. Payne is better known for his discovery of songs in Humpback Whales. But before he went to whales, he conducted a set of brilliant studies on barn owls, exploring the precision of their strikes and the exact sensory cues they use to locate their prey. In one experiment, Payne blocked all light from a room so it was pitch black, and placed an owl on a perch in a corner. He covered the floor with leaves and then dragged a mouse-size wad of crumpled paper through the leaves. The owl tried to strike the rustling paper. The bird wasn't using sight or smell or body heat in targeting its prey. "The paper wad and the leaves through which it was dragged were at the same temperature," wrote Payne. "Therefore, the owl could not have located it by any infra-red contrast with its surroundings. The paper wad put out no mouse-like odour, so homing on scent would be of no value. Because the lights were out, the owl could not see the wad. . . . The only possibility left, as I see it, is that the owl was orienting acoustically to the sounds."

Just to be sure, Payne tried blocking the owl's ear with a small cotton plug, first one ear, then the other. He let a mouse loose to scurry in the leaves. "In both cases the owl flew, in darkness, directly towards the mouse but landed about eighteen inches short of it," wrote Payne. "After each trial the cotton was removed, and the owl was allowed to try in total darkness to catch the same mouse it had just missed. In both cases the owl then struck successfully."

Payne also filmed owls targeting prey on the wing in complete darkness. The results were mind-boggling. When a mouse shifted direction, the owl would twirl its head toward the creature and adjust its attack midflight.

How in the world does a bird do this?

With a head designed for listening, like Percy's. The flat, gray facial disk of a Great Gray Owl is like one huge external ear, a feathered satellite dish for collecting sound. Not all owls have the big, pronounced facial disks of Great Grays, Boreal Owls, and barn owls. It's smaller in owls that rely less on sound for hunting—Great Horned Owls, Little Owls, pygmy owls. And in some species, like fish owls, it's dramatically reduced. This makes sense. Streams are noisy, water is noisy, and sound reflects off the air-water surface. So, presumably, a hunting owl can't hear a fish. But fish owl expert Jonathan Slaght thinks the birds may be using sound more than we think. He showed me a photo of a Blakiston's Fish Owl on a riverbank in which the bird "really looks like it's using its facial disk," he says. "So I think these 'owly' features are diminished but not gone."

The facial disk in owls that hunt primarily by sound is outlined with a ruff, or ring of stiff interlocking feathers that capture sound waves and channel them toward the ears, like people cupping their

Great Gray Owl facial disk

hands around their ears. Feathers in the back of the disk direct high-pitched sounds toward the ears, so the owl hears less noise from its surroundings and can focus on prey cues. "The diversity of feathers on a Great Gray Owl's facial disk is just phenomenal," says Jim Duncan, an expert on the species. "Seven or eight different kinds. The ones you see are very loose and filamentous, and sound travels through them quite easily. And then there are these curved solid feathers that form the back of the facial disk and act like the parabolic reflector part of the disk. The curve probably reflects the optimal angle for sounds hitting the disk to be directed into the ear cavities." Percy can even change the shape of the disk by using muscles at the base of the feathers, shifting from a resting state to the alertness of an active hunt. It's remarkable to watch an owl do this, adjust its facial disk when it hears something interesting. It's like the disk itself is a kind of aperture, an "eye," that opens wide to let in more sound and bounce it toward the ears.

The use of the term *eared* in the common names of some owls is confusing. Long-eared and Short-eared Owls have tufts of feathers on the tops of their heads, called "plumicorns" (from the Latin for "feathered horn"), which look a lot like mammal ears. But these tufts have nothing to do with hearing and everything to do with camouflage and, sometimes, display.

An owl's actual ears are just openings in each side of the head, well covered with specialized feathers that allow sound to pass through. Their size varies from species to species, depending not just on whether they hunt in daylight or darkness, but on the general invisibility of their prey. The night-hunting Long-eared Owl, which feeds heavily on small rodents, does in fact have long ears in addition to long tufts, with ear slits that run from the top of its head all the way down to its jaw. Barred Owls and Boreal Owls, which are mostly nocturnal, also have large ear openings. But so do pygmy owls, which often hunt by day, because the little rodents they prey on are often hidden in dense grassland and must be sussed out by sound.

And Percy's prey? In the wild anyway, it's often buried deep in snow, which not only obscures everything visually but also creates what's called an "acoustic mirage," distorting the location of sounds beneath it and making it harder for a bird to pinpoint its prey. As we'll see, Great Grays have evolved some truly spectacular strategies to deal with the challenge.

In any animal ear, a little sliver of tissue called the "cochlea" collaborates with the brain in the hard work of hearing. The cochlea contains hair cells sensitive to the vibrations of sound, and its length in an animal is a pretty good measure of hearing ability. In most owls, the cochlea is enormous relative to body size and contains huge numbers of hair cells compared with other birds. The cochlea of a barn owl, for instance, is massive. "It's the race car equivalent of the bird inner ear," says Christine Köppl, who studies barn owls at the University of Oldenburg in Germany. In her talks, Köppl shows a slide comparing the cochlea of a barn owl with a range of other bird species—blackbirds, jays, buzzards, and hawks. The owl's cochlea is easily three or four times as long as those of other birds, giving it extraordinarily acute hearing.

An owl's auditory system shares with other birds another superpower we mammals don't possess: it doesn't age. To see whether the hearing of barn owls changes over time, Köppl collaborated with two colleagues, Ulrike Langemann and Georg Klump. The scientists trained seven owls varying in age to fly from one perch to another to receive a treat in response to an auditory signal. Then they separated the birds into "young" and "old" age groups and tested their hearing by changing the tones, moving them up and down the frequency scale. The team found no age-related hearing loss at all between the young barn owls and the older ones. In fact, the star of the study, a twenty-three-year-old Methuselah of a bird named Weiss, could hear the full range of tones just as well as the two-year-old birds in the study. This suggests that owls, like other birds, have the capacity to regenerate their hair cells, keeping their hearing keen throughout life.

We mammals are not so fortunate. Growing old as a human or mouse or chinchilla brings with it age-related hearing loss, especially in the range of high-frequency sounds. In our ears, damaged hair cells are not replaced as they are in birds, and we can only envy the regenerative powers of owl ears.

A Great Gray Owl is listening, always listening. Its head rotates to glean the source of a sound. Its ears are so acutely tuned, it can discern the faint footfall of a shrew in the forest, the wingbeat of a Canada Jay, the muffled rustle of a vole tunneling deep beneath the snow. It will fly to the spot, hover over it, head facing down toward the sound, then just before impact thrust its legs forward and punch through snow more than a foot and a half deep to seize its prey.

To be able to hunt by sound alone, owls need not only supersensitive ears but also the ability to locate the source of a faint noise in three-dimensional space—sometimes from a distance and sometimes through a thick layer of snow, soil, or leaves. The late Masakazu (Mark) Konishi tackled the problem of how an owl might do this.

Konishi died in 2020. A year later, on the anniversary of his birthday, a large group of researchers—peers and graduate students—gathered for a virtual celebration to honor the scientist and the man and to bring to light new research inspired by his work. The titles of the talks reflected the feeling of awe they shared with Konishi: "The Amazing Barn Owl Cochlea," "The Owl's Amazing Midbrain," "The Amazing Nucleus Laminaris."

When Konishi heard Roger Payne report that a barn owl can catch a mouse relying only on sound, he wanted to understand exactly how a bird could do this. How can an owl track its prey in complete darkness? How can it work out exactly where a sound is coming from? What sort of brain circuitry allows for that? Konishi knew that facial disks helped in the task, and also the asymmetry of ears—at least in certain owl species.

Some owls, such as Great Horned Owls and Eastern Screech Owls, have ears placed at about the same level on both sides of their heads like most animals do. But others—barn owls, Northern Saw-whet Owls, and Great Gray Owls—which rely heavily on sound for hunting, have one ear hole higher on one side of the head than the other. The asymmetry of Percy's ears is stunning. Beneath that mass of soft feathers, the left ear sits just below eye level, the right, slightly above. To accurately locate prey, Percy compares the sounds arriving at each ear, how loud they are, and which ear detects them first. Percy's right ear is more sensitive to sounds coming from above the midline of his face, while the left ear is more sensitive to sounds coming from below. The difference in the time of arrival of sound waves between his two ears, known as the interaural time difference, helps Percy gauge the exact azimuth (or horizontal location) of a sound. The difference in loudness between his two ears helps him to determine the sound's elevation. Where azimuth and elevation intersect is where he directs his strike. Species like Great Gray Owls, barn owls, and Northern Saw-whet Owls can locate sounds within just two or three degrees.

There's more to it. Tracking prey precisely takes two ears, and their asymmetrical arrangement helps. But in the end, it's the brain that locates sounds in space in a most ingenious way.

By the time Konishi moved from Princeton to Caltech in 1975, he had twenty-one barn owls trained to strike at loudspeakers producing all kinds of sounds, including one owl named Roger, after Roger Payne. (It should be noted that Roger the owl turned out to be female; at one point, "he" laid an egg.) Roger starred in so many publications that the researchers celebrating Konishi thought he might be among the most famous published animals, rivaling Alex the African Gray Parrot, who, together with Harvard scientist Irene Pepperberg, taught the world so much about bird brains and intelligence.

Konishi's research got a boost when a Caltech machinist famous for working on the Viking lander in the first Mars mission designed and built some fancy equipment for his owl studies—an ingenious light-rail, arranged in a semicircle. Attached to the rail was a small remote control loudspeaker that could travel around the head of an owl at a constant distance in both horizontal and vertical directions. With the help of this space-based gadget, Konishi and his doctoral student Eric Knudsen made a remarkable discovery. Certain auditory neurons in an owl's brain respond only when a sound is coming from a particular location. By comparing the responses to sound by neurons in the cochlea of both ears, the brain builds a kind of multidimensional map of auditory space. This allows owls to fix the location of prey with speed and precision.

This was a surprise. Animals have brain maps for vision and touch, but these are built from visual images and touch receptors that map onto the brain through direct point-to-point projections. With ears, it's entirely different. The brain compares information received from each ear about the timing and intensity of a sound and then translates the differences into a unified perception of a single sound issuing from a specific region of space. The resulting auditory map allows owls to "see" the world in two dimensions with their ears.

This proved to be a big leap toward understanding how the brain of any animal, including humans, learns to grasp its environment through sound. Think of it. Standing in a forest, you hear the crack of a falling branch or the rustle of a deer's step in the dry leaves. Your brain calculates the time and intensity of sound to determine where it's coming from. Owls do this task with incredible speed and accuracy. Each cochlea in the owl provides the brain with the precise timing of the sound reaching that ear within twenty *micro*seconds. This determines how accurately the brain can calculate the interaural time difference, which in turn determines the accuracy of the localization of a sound in the azimuth. "The precision in microseconds provided by the owl cochlea is better than in any other animal that has been tested," says Köppl. "We have big heads, so the interaural time differences are larger, making the task for cochlea and brain easier. In a nutshell, it is the combination of a small head and very precise localization that makes the owl unique."

And here's a finding to drop the jaw. José Luis Peña, a neuroscientist at the Albert Einstein College of Medicine, and his collaborators have discovered that the sound localization system in a barn owl's brain performs sophisticated mathematical computations to execute this pinpointing of prey. The space-specific neurons in the owl's specialized auditory brain do advanced math when they transmit their information, not just adding and multiplying incoming signals but averaging them and using a statistical method called "Bayesian inference," which involves updating as more information becomes available.

All of this calculating in less than the blink of an eye. I know, it's mind-blowing.

It's not only an owl's hearing that's keen. Owls have exceptional vision, too, and research on the way the two senses work together has yielded some fascinating insights on owls and on human babies, too.

When I edge closer to Percy, he looks as if he would roll his eyes at me, if only he could. The big eyes of owls are tubular, rigid, and locked in their sockets in a forward gaze. Not the norm in birds. Birds typically have oval or disk-shaped eyes on the sides of their heads.

Why would an owl have forward-facing eyes?

Graham Martin, who has studied avian vision for more than fifty years, has argued that it's for one very simple reason. Their size. Percy stands around two feet tall and weighs only two and a half pounds or so, but his eyes account for about 3 percent of his body weight. My orbs weigh in at just .0003 percent of my poundage. If my eyes were in similar proportion to my body as Percy's is to his, they would be about the size of an orange and weigh almost four pounds. The eyes of owls are forward facing because they're so big, argues Martin, and an owl's skull is so small and crowded with big, elaborate hearing structures that there's nowhere else for the orbs to fit. Indeed, you can see the side of an owl's eye through its ear opening, "indicating that eyes and ears are very tightly packed within the skull," Martin writes. "Where else can the eyes be positioned?"

That may be so, but forward-facing eyes also give owls a vital gift for hunting: binocular vision. Not as much as we have, but plenty compared with most other birds. For a sparrow or titmouse pursued by an owl, eyes on the side offer a wide field of view, the better to see that predator coming. Owls have a narrower total field of view, but their binocular vision gives them an enhanced ability to determine their direction of travel and the time required to reach a target—all big advantages in zeroing in on prey, especially if it must be caught with split-second timing. However, having a narrow overall field of view has consequences. Stand near an owl, and it may bob and circle and weave its head from side to side, forward and back, up and down, sometimes torquing it until it's nearly upside down. The bird is trying to get a good look at you.

That Percy's eyes are fixed in a forward gaze means that the only way he can follow my movements is by swiveling his head. Fortunately, he's good at that. While it's a myth that owls can rotate their heads full circle from a starting point facing forward, some species, like Great Grays and barn owls, can turn their heads almost three quarters of the way around, 270 degrees—three times the twisting flexibility humans possess. Owls have exactly twice as many neck vertebrae as humans do, giving them that much more flexibility. Other birds have the same number of neck vertebrae and are capable of twisting 180 degrees or more to preen. But their necks are not buried in feathers like an owl's is, so it's easier to detect the bending and twisting as they "crane" to see behind them. That an owl's neck can move swiftly and smoothly through those 270 degrees of rotation is due to some clever adaptations, a loose S shape that gives it flexibility, and a system of bones and blood vessels that minimizes disruption of blood flow through the neck to the eye and the brain when the head rotates.

In 2016, scientists exploring how vision evolved in birds turned up a secret about owls. The team studied 120 vision genes in 26 different bird species, from owls and hoopoes to falcons and woodpeckers.

Owls, it turned out, were the queens of visual adaptation, with more modifications to vision-related genes than any other group of birds.

"The nocturnality of owls, unusual within birds, has favored an exceptional visual system that is highly tuned for hunting at night," write the scientists. Over evolutionary time, it seems, owls made a kind of sensory trade-off. They lost some of the genes involved in daylight and color vision. But their genes for nocturnal vision were enhanced and refined. Peregrine Falcons and other raptors may have sharper vision in daylight, allowing them to discriminate fine detail at greater distances, but the big tubular eyes of owls admit more light and have more cells that process photons, giving them visual acuity even in the dimmest conditions. Most birds have a retina dominated by cones, cells that work best in bright light to help with color detection. The retinas of owls are packed with rods, which are much more sensitive to light and movement. Night-hunting owls like Percy have about 93 percent rods versus 7 percent cones, giving them about one hundred times the light sensitivity of a pigeon. Their night vision is better than ours, though not as keen as a cat's.

Owls may have lost some genes for daylight vision, but they held on to something rather extraordinary that most day-active birds have—sensitivity to ultraviolet light. Diurnal birds have four kinds of color-sensitive cones in their retinas, including one that responds to ultraviolet light, which allows some species to perceive another whole dimension of color, millions of hues we can't see. Now new studies are showing that some owls have retinal *rods* that are sensitive to ultraviolet light. Why would a bird active mostly at night, when color is largely imperceptible, need UV sensitivity? There are good reasons, as we'll see later, having to do with feeding chicks and repelling rivals.

For owl eyes, it's all about light. An owl's pupils can swell to nearly the entire size of the eye, letting in about twice as much light as human pupils. And its pupils dilate even more if it hears a new noise—a link between sight and sound that enhances its hunting skills. (Percy's eyes did indeed widen at the sound of that zookeeper dipping his hand in the bowl of mice.)

When neuroscientist Avinash Singh Bala stumbled on this insight some years ago, it was completely unexpected—and, it turns out, useful to our understanding of hearing in human babies. Bala was training barn owls to respond to different sounds for a study aimed at understanding how human brains process sound. While he was setting up the experiment, he noticed that the owls' eyes would dilate in response to an oddball noise—a door slamming or something dropping on a desk. Later, he realized that humans, too, have this involuntary pupil response and that it could be used to measure hearing not just in owls but also in babies. Because babies can't tell you what they hear, diagnosing hearing issues is challenging. The discovery that their pupils, like those of owls, respond to a new sound, including slight variations in the volume or content of a word like *bah* and *pah*, led to a new diagnostic test for hearing loss in infants. It's a telling example of how basic owl science has boosted human medicine—and how an owl's eyes and ears work together.

No other birds use vision and hearing in such a highly coordinated way to detect and catch prey items, says Graham Martin. "In most birds, it's likely that vision and hearing serve different functions in guiding their behavior, but in owls the two senses come together to locate an object during prey capture. The highest hearing acuity of owls lies directly in front of the head in the binocular field of view. If they can see where a sound is coming from, in addition to hearing it, they can use vision to estimate direction and time to contact as they pounce."

A team of Dutch scientists studying brain anatomy in owls recently turned up another eye-ear connection: part of the hearing nerve that goes to the brain branches off to the owl's optical center as well. "This indicates that part of the sound information owls get through hearing is processed in the brain's visual center, so owls may actually get a visual image of what they hear," speculates Kas Koenraads, a morphologist and ecologist from the Netherlands conducting the research. "We don't know what this means for an owl. It could be that if an owl hears something moving in a dark, forested environment, it gets some kind of visual indication of where the audio cues are coming from, like an illuminated dot of light in the dark forest. It would be really cool if it works like this," he says. "Based on morphological characteristics, it's a possibility. But we're probably never going to find out because we can't crawl into an owl's head."

Imagine if we could.

Imagine if just for a moment we could notice what an owl notices, hear every small sound in the woods as if we were, say, a Great Gray Owl like Percy gathering food for his mate or his chicks in the pitch of night—not from that zookeeper's bowl but from a dense, entangled thicket of grass and sedges or under deep snow. From his perch, Percy scans with those intense light-gathering eyes, probing the dark, collecting whatever scarce and scattered photons may exist and taking them into his widened pupils. He swivels his head to tune in to that faintest of rustles, focusing the vortex of his attention on the dim twitching on the ground or beneath the snow. His ears lock in on his prey. Then, on wings soft as a warm breeze, he bears down on the creature with such swiftness and silence it never sees what's coming.

Much of this exquisite equipment in owls is aimed at locating hard-to-spot prey making impossibly faint sounds that come and go in an instant. When we listen for feeble noises, we quiet ourselves as much as possible. Consider how silent a hunting owl must be. I think of Percy in flight, his massive body approaching me on that railing. His wings almost brushed my ear, yet no sound was audible, not a single rustle of air in feathers or breathy whoosh.

An owl's quiet flight: it's one of the great wonders of the bird world, only beginning to yield its mysteries.

QUIET FLIGHT

If you have ever seen an owl wet, you can fully appreciate their feathers. Not long ago, a wildlife rehabilitation clinic in New Zealand posted an image on social media of a Morepork in its care, before and after a bath. The little brown owl (also called by its Maori name, "ruru," for its two-note call) was all attitude and fluff before its bath. But the vets had noticed bacteria on its skin, so they gave it a special soaking that temporarily but completely flattened its puffy plumage, leaving it scraggly and bedraggled. The contrast was so stark that the image racked up 37,000 Likes in just two days. What *was* this sticklike creature?

The greatness of a Great Gray Owl is considerably diminished when you feel beneath the feathers to its delicate skull and bones. It is only the full four inches of feathers surrounding his skull that gives Percy the illusion of hugeness.

While they may not be waterproof, owl feathers have evolved ingenious adaptations for camouflage and quiet flight over millions of years. Their color alone is a wonder of specialization, dominated by natural hues, shades of brown, cream, and gray. The dark surfaces of their feather vanes (the webby part) are saturated with the pigment melanin, which gives them not only extra strength and hardness but also resistance to abrasion and to feather-degrading bacteria and parasites.

Great Horned Owl after rain

With all the benefits of dark plumage, why aren't owls uniformly dark colored? Because it takes extra material and energy to produce dark feather parts, including deposits of minerals like calcium, cadmium, and zinc. (It's not easy for owls to take in calcium—they digest the bones of their prey less efficiently than other birds of prey.) Also, darker colored feathers are heavier. The lighter colored parts of feathers weigh up to 5 percent less than the adjacent dark portions. So owls are thrifty in their dark coloring, with strategically placed barring and spotting on the upper surfaces of their wings and back and paler plumage on their underwings and bellies. Some owls, particularly forest species, have dark, strongly saturated feathers at the front edges of their wings and their wing tips, to provide strength against collisions and the wear and tear of flying through vegetation. Big owls, like barn owls, tend to have strong feather shafts and ribs of dark barring against a lighter, paler background (which costs less to grow), whereas smaller

Barn owl wing showing feather pigmentation

owls have weaker feather shafts, reinforced by darker vanes, patterned with pale ovals and spots (again, to save energy costs and weight).

Like other birds, owls molt to renew their plumage, regularly shedding feathers that are old or worn from rubbing together in flight or damaged by collisions with branches or grass or passing through the narrow opening of a tree hollow or cavity. Especially vulnerable to wear from adjacent feathers are the lighter colored parts of vanes that are not reinforced by dark pigment. Molting renews a bird's plumage and keeps it in top condition for efficient flight and thermal insulation.

Owls generally have more feathers than most other birds. Not long ago, when David Johnson and a team of volunteers counted every feather on a dead female Great Horned Owl—a labor involving forty-six hours of work—they came up with 12,230 individual feathers. Eagles and most other birds of prey have about half as many. The Long-eared Owl has four times the number of feathers of other similar-size non-owl birds. The feathers tend to be most dense around an owl's face. The head of the owl Johnson examined held some 32 percent of all feathers, close to 4,000—more than any other part of the body—but represented only 7 percent of the total weight of the feathers. Beneath the grayish lightweight feathers in the facial disk of a Great Gray are beautiful orangish-yellow feathers. Jim Duncan speculates that Great Grays were perhaps once more colorful, and evolution has moved all the exposed feathers toward neutral gray, brown, and white for camouflage. Some owls even have bristlelike feathers around the beak that act as touch receptors. And some, like Snowy Owls, have heavily feathered feet.

Because feathers tend to make a lot of noise, the quiet flight of owls is all the more remarkable. I once witnessed a demonstration of a free-flying hawk by a master falconer. The falconer asked a group of about thirty of us to lie side by side in a grassy field. When the giggling subsided, his assistant stood at one end of the field of bodies with a Harris's Hawk on her arm. The falconer stood at the other end and told us to close our eyes. As I lay there in the grass with eyes shut, heart beating in anticipation, it was easy to imagine what it's like to be prey, alert to any sound, any whiff of air or motion. The assistant released the bird. Suddenly, there was a loud whir of wingbeats and, for an instant, as the bird swooped low over us on its way to the falconer, the whoosh of wings and turbulent air.

All birds make sound in flight. Noise arises from the drag of air over a bird's body, from the vortices, or turbulent whirls of wind in a bird's wake that generate sound waves, from air squeezing through the slits in feathers, and from feathers rubbing together. With every flap, bird wings rustle and flutter, whistle, hum, and whish. They drum

Barn owl in flight

and snap and clap. But the sounds that many owls make when they fly are so faint that they're below the threshold of human hearing. Measurements in a lab comparing a Harris's Hawk and a European Kestrel on the one hand, and a barn owl and a Eurasian Eagle Owl on the other, showed that the wings of the two owls generated noise five to ten decibels less than the wings of the other species.

Despite decades of study, an owl's furtive flight is a feat of biomechanical stealth that still challenges biologists and engineers, and it's one of those superpowers that make owls such a pinnacle of evolution.

Google "owl silent flight" and you'll find a dramatic video of an experiment set up by BBC Earth comparing the flight noise of a pigeon, a Peregrine Falcon, and a barn owl. The crew filmed all three birds with a slow-motion camera as they swoop down a flight path outfitted with six super-sensitive microphones. It's an eerie setup, a bird flying in the dark, with a spectrogram of the sound it makes hovering in the black. The differences are dramatic. The flapping of pigeon and falcon show up on the spectrogram as big spikes of sound. The owl's acoustic wave is flat.

Christopher Clark believes that the clip may have been doctored in some way. "Owls *do* make sound when they fly," he says.

Clark would know. He runs the Animal Aeroacoustics Lab at the University of California, Riverside, which studies the sounds animals make in flight. He and his graduate student Krista Le Piane conducted their own experiment with barn owls flying less than half a meter above a microphone. The owl's flapping shows up as a low-frequency waveform, *very* low frequency, but it's there. (Also faintly audible is a higher frequency component of sound, which Clark thinks is made by feathers.)

Still, the BBC's point is well taken. Owls may not be silent fliers, but they are nearly so. In part this is because owls have low wing loading—their wings are big in relation to their bodies—so their flight is buoyant and slow, as slow as five miles per hour for a big bird like a barn owl, which makes it quieter. (Owls need to fly slowly to stalk prey in open fields and to navigate through trees and other obstacles in forests.) But it's the marvelous and unique feathers and structure of owls' wings that really hush their flight.

Chris Clark is fascinated by feathers, especially how birds make sound with them—and, in the case of owls, how they suppress the sound feathers naturally make. He has recently taken a deep dive into the ways owl feathers have evolved to squelch sound, giving these birds their extraordinary gift of stealth. Some of his findings confirm what we already know. Others turn our old understanding topsy-turvy and pose intriguing new riddles.

An avid birdwatcher since childhood, Clark was captivated by the acrobatic flight of the Anna's Hummingbirds he saw as a teenager at the University of Washington arboretum in Seattle. He began his career studying flight biomechanics, specifically the three-dimensional kinematics of hummingbird flight, the way these tiny birds hold their bodies and shift their wings in forward flight.

"I had no interest in sound, and it never occurred to me that noise produced during flight might be important," he says. But watching Anna's Hummingbirds changed all that. In 2008, Clark discovered a wonder in the hummingbird's courtship performance: during its breathtaking dive display to woo potential mates, the bird *sings* with its feathers—or, as he writes in his dissertation paper, "chirps with its tail." In that paper, Clark demonstrates that the loud chirp the male hummingbird makes at the nadir of its dive, just when the bird is directly over the female, is not a vocalization as was previously believed but a curious burst of squeaky mechanical sound produced by an exquisitely timed high-speed flutter of its specialized tail feathers. "I spent the next decade or so measuring how differences in shapes of the feathers of different species of hummingbirds affect the sounds they produce," he says. "I was asking the question 'How do some birds generate extra sound when they fly?' And it occurred to me that quiet flight is kind of the opposite side of the same coin, and equally ingenious.

"Feathers are amazing structures and probably one of the reasons birds are so incredibly successful," he says. "The keratin that makes up feathers can handle higher strain rates than aluminum, and that means feathers are extremely flexible and can bend to a greater angle before they start to have fatigue or other problems."

But feathers have other qualities that might be a disadvantage for a bird that hunts by stealth. "Feathers are autonomous and rough with barbs and barbules at a submillimeter scale, so they tend to produce frictional sound in flight," Clark explains. "Rub together two tail feathers of a Red-tailed Hawk and you get a fair amount of frictional sound, like two pieces of Velcro unzipping or a piece of sandpaper rubbing against a surface." Because most feathers are noisy, most birds produce what Clark calls an "audible signature" with every flap of their wings—just as that Peregrine Falcon did in the BBC film and the Harris's Hawk did as it flew over us in the field. Vultures and hornbills make noise even when they're gliding. Not owls.

As early as 1934, Robert Rule Graham, a British pilot and bird lover, identified three features that suppress sound in owl flight. First, he observed an unusual feature known as a comb, a row of fine hairlike bristles that extend forward along the leading edge of the wing (where it meets the oncoming air). He also noted a belt of ragged wispy vane fringes on the wing's trailing edge (its rear edge) and, finally, a soft layer of velvet coating the whole wing.

Graham was mostly right. These three mechanisms are key to quiet flight. But in the decades since his observations, a bevy of biologists and engineers have refined his interpretations, analyzing the various wing features he described and even using owl-inspired designs to fashion quieter airplane wings and fan and wind turbine blades.

In most birds, air flowing over the wing surface produces turbulence, air eddies that make noise. One team of researchers studying the leading-edge comb of a barn owl's wing discovered that when the airflow hits the comb-like serrations, they break up the turbulence, effectively suppressing that swoosh sound I heard from the Harris's Hawk. When an engineer tested the number of "teeth" in the comb on the wing of a Eurasian Eagle Owl, he found it has an effective number to reduce turbulence, twenty-eight per inch. The comb also works to quiet otherwise noisy flow at the wing tip, especially just before an owl lands, when it's close to a stall during the final stage of an attack. These features likely explain the findings of engineers who recently performed experiments with flying Australian Boobook Owls, creating computer simulations of the airflow around the wings of a flapping owl: the air shed from the owl's wing is broken up, not "organized" the way it is in other birds with noisier flight. Another research team that recently analyzed the primary wing feathers of owls found they have soft, elastic tips, quashing noise that might arise around a stiff-edged wing.

But in Chris Clark's view, an owl's quiet flight primarily comes down to a remarkable ability to reduce frictional noise between feathers.

Pick up a pair of owl feathers and rub them together, and you won't hear much. That's because the feathers are coated with a fine layer of plush fibers called "pennula," which shroud sound and give owl wings that soft, velvety feel that Graham noted. An owl's wing feathers separate slightly from one another in flight, so air flows over each feather, with the pennula providing a gap between adjacent feathers so there's none of the friction or rubbing there is in most birds. The wispy vane fringes at the tips of both wings and tail also help to prevent wind eddies. Collectively, the serrated edge, the pennula, and the wispy tips unite each of the feathers into a single soft surface without sharp, noisy edges.

Not every owl is plush and soundless on the wing. The tool kit for quiet flight varies from species to species. Owls that depend less on hearing when they hunt, such as Mountain Pygmy Owls, have noisier flight. The Great Gray Owl has the most extreme traits related to quiet flight. It has the thickest velvet, one of the longest leading-edge combs, and possibly the most extensive vane fringes of any owl.

Why would the Great Gray be such a "quiet flight extremist"?

In 2022, Chris Clark and Great Gray expert Jim Duncan conducted an ingenious experiment that suggests the answer goes back to the way this owl hunts voles beneath snow.

Duncan had always been intrigued by the Great Gray's ability to locate prey under snow up to a foot and a half deep. Early in his research on these owls in Manitoba, Canada, he noticed something unusual: as winter progressed, the Great Grays were getting fatter. "Winter in Manitoba is no cake walk," he says. Many species of wildlife suffer.

Great Gray Owl hovering

Up to 40 percent of the population of White-tailed Deer can be killed in a single winter. "Great Grays seem to be one of the few species that just love winter. So it was this burning question in my mind, What allows them to be such successful hunters in the snow?"

In Manitoba, Clark and Duncan tested the impact of snow cover on the recorded sounds of voles digging around beneath it, using a special acoustic camera that creates images of sound.

Anyone who knows the hush that falls over a snowy landscape is aware that snow absorbs and muffles sound. But as the scientists discovered with the sound camera, snow also creates an "acoustic mirage," refracting sounds from beneath it so they're displaced. It's like the way water refracts light, so that when you're looking at a fish at an angle from a boat, where you see the fish is not where the fish actually is because the water is bending the light. Snow is frozen water, and it bends sound. The only way for an owl to detect the true location of a vole's snuffling under snow is to hover directly above it. And that is exactly what a Great Gray does. Most owls fly directly toward prey. Great Grays hunting in snowy areas fly to a spot above the prey and then hover for up to ten seconds before plunging straight down. (This is the same tactic Ospreys and kingfishers use to cope with the effects of light refraction in water, hovering directly above their target and then plummeting.)

And the extra-thick velvet and long combs and vane fringes of the Great Gray? They work specifically to dampen the flight sounds made during this hovering.

Why is quiet flight so vital to owls? Is it to avoid making noise that would interfere with an owl's own hearing of its prey—what Clark calls the "owl ear" hypothesis? Or is to prevent prey, that cowering mouse, from hearing the owl's approach—the "mouse ear" hypothesis? That is, do owls need silent flight so they can locate prey by sound while in flight? Or for stealth, to allow them to approach their prey undetected?

One piece of evidence Clark has found for the "owl ear" hypothesis: owls with large facial disks also have the longest comb serrations, suggesting that species that rely more on hearing for locating prey also have quieter flight. But he would be the first to say that these "owl ear" and "mouse ear" functions are not mutually exclusive. "In science, we love putting things into binary opposition, but the truth is both functions likely play a part."

To parse whether owls evolved quiet flight to reduce the noise they themselves make, the better to hear their prey, or to sneak up on that prey with more stealth, the better to ambush it—or, as is most likely, both—"you really have to know what owls hear of their own flight," says Clark. It's that big challenge of getting inside an owl's head. "To what degree do they go into stealth mode, changing their own kinematics"—the properties of their motion—"in order to fly more quietly? To what degree do they change their behavior if they're having a hard time hearing their prey?

"And also, what does their *prey* hear? What does a mouse hear when an owl swoops down on it?"

Clark has ideas for a range of clever experiments to try to unpack these questions. He has considered putting a microphone in a little backpack on an owl to record what the owl hears when it flies and when it hunts. "But the technology isn't quite there yet," he says.

A few months before we spoke, Clark read a scientific paper in which the authors had put small data loggers on spotted owls to record the sounds they made. "I got all excited and I emailed the authors saying, 'Oh, this is great! This is amazing. I didn't know they made data loggers that small,'" he recalls. "Then they sent me some of the recordings, and they were *so* disappointing, terrible quality. You can hear the spotted owl hooting. You can hear when it grabs a rodent of some sort. You can hear the death scream of the rodent. But the microphones have such low sensitivity, there's no way of measuring the wing sounds the owl produces when it flies.

"As far as I can tell, the technology just isn't there yet for actually measuring what the owl hears as it approaches prey," he says. "So really, for this experiment, you would want little microphones *inside the ears* of the owl to record what the owl is hearing as a function of how it flies. Once you got that, you could go about trying to figure out: Do the owls adjust their kinematics in response to the specific sounds to make their flight stealthier?"

Recording from the owl's perspective is really difficult, says Clark. Recording from the prey's perspective may be easier. He's working on that now, using domestic mice from a pet shop. "Nobody has put a microphone next to a mouse and recorded the mouse's acoustic experience as it's being attacked by an owl," he says. This is important to tease apart the "owl ear" from the "mouse ear" hypothesis. But even this strategy is fraught with problems and a good example of just how challenging these studies can be.

Clark has had an experience with this approach—well, two, actually—and neither went well.

"We have a field station out in the desert," he explains. "Deserts are phenomenally quiet environments, especially in the fall and winter. And so I went out there in September and put out a mouse in a small cage, with a microphone next to it. Then I put some dry leaves inside the cage with it. When you want to catch an owl, this is what you do so that the mouse walking around in the cage makes a sound that the owl can hear." Then he rolled out his sleeping bag next to the cage and went to sleep. At five o'clock in the morning, he suddenly woke up and found a Great Horned Owl right there next to him, but it saw him and flew off. "Checking the sound recording later, all it captured was the mouse walking on dry leaves," he says, "this really loud *crinkle crinkle crinkle, crinkle crinkle crinkle*, and then the *wham* of the owl hitting the trap. Because the mouse itself was making so much noise, I didn't get a recording of what I wanted—the flight sounds the owl might make, as heard by the mouse. I just got a recording of a mouse walking on leaves."

How to get around that problem? At a golf course near his house, Clark tried luring the owls that frequent the place not with an actual mouse but with a speaker playing the sound of a mouse rustling around in leaves. "Within like one minute, a barn owl showed up and landed on the tree right above me," he says. But the owl seemed to realize that something was wrong. "It just sat there for twenty minutes looking at the speaker. At one point, it flew over the speaker and past it, and then hovered over some tall grass, like it was looking for the source of the sound, not where the speaker was but *behind* it."

Clark thinks he had the volume turned up too high. "To the owl it probably sounded like the world's largest mouse crawling around in the world's largest leaves."

Jim Duncan is approaching the question of what an owl hears from another angle. He is planning to investigate just how the big facial disks of owls like Percy act to augment the volume of sounds. He intends to place tiny microphones in the ears of owls—dead owls—to measure and quantify the sound magnification effects the facial disk has on weak sounds emanating from the ground. So far, he has five owl heads in his freezer, most of them from owls killed in vehicle collisions. "It sounds gruesome," says Duncan, but it could shed fascinating new light on what an owl hears.

The extraordinary effort of these scientists to hear what an owl hears—and what its prey hears—is a good example of the challenge of understanding owls and the tenacity and creativity it takes to unlock their secrets.

Three

Owling

Studying the World's Most Enigmatic Birds

FISHING FOR OWLS

Owls are hard to study in the wild for some of the same reasons we love them. They're quiet, wary, secretive, and often elusive. They're masters of camouflage—streaked like grasses; mottled, speckled, and striped like tree bark; pale like snow—to befuddle the eye of both predators and prey. They wear the look of the land around them to meld into it, a strategy known as crypsis. The Flammulated Owl sports patches of reddish-orange plumage that obliterates its form, making it blend almost seamlessly into the mottled trunks of Ponderosa Pines. The ear tufts of some species disrupt the round identifiable shape of an owl head and help it blend in with its woody surroundings. The Eastern Screech Owl fills the tree cavity it roosts in, its brown and gray feathers merging with the surrounding tree. Even a bird as big as the Great Gray Owl can vanish against the bark of a larch tree, the very portrait of ambiguity. To complete their disguise, owls act the part of the tree or field that's concealing them, staying utterly still, letting their tufts sway in the wind in a way that can mimic branches or grass.

Owl researchers have had to devise a host of inventive ways to scout out their subjects. Listening for their calls, for instance, and trying to elicit these vocalizations either by hooting themselves or with playback, broadcasting recorded territorial calls with amplification devices and waiting for a response. Owls are extremely territorial, so if they're present, they often respond with their own hoots. In conducting these vocalization surveys, it's standard protocol to avoid playing the calls of the biggest owls first. Big owls eat little owls. Play the booming call of a Great Horned Owl first, and the little owls go quiet. So researchers often start with the toots of smaller species. Or if they're trying to locate just one species, they play only that bird's territorial calls.

Finding owls this way is labor intensive and also "a little bit art, a little bit science," says Jennifer Hartman, who has surveyed extensively for Northern Spotted Owls with a research team in the Pacific Northwest. "You have to listen really, really intently. Sometimes the owl is distant, and all you hear is a faint hoot or whistle, and you mark that spot with a compass and then you go to three or four other points to try to narrow in on the spot where the owl is hooting," she says. "It's a secret skill, and the more you do it, the better you become. If we were lucky enough to find an owl, my heart always skipped a beat. It was the most magical thing. Over the three or four years I did this, I found hundreds of owls, but it was still magic to me every time we found one, because they're so well camouflaged and so shy."

They are indeed camouflaged and shy. And sometimes uncharacteristically quiet, too. In their efforts over the years to get accurate counts of Northern Spotted Owls, Hartman's team ran into a glitch with the usual vocalization survey method. Barred Owls had invaded the home range of the spotted owls, and the presence of the aggressive invaders suppressed the smaller owls' own hooting. If the spotted owls did vocalize, the Barred Owls would attack them, sometimes with lethal force. So the smaller birds started to go silent, and the traditional

Northern Spotted Owl

method of hooting to find Northern Spotted Owls stopped working. "The land managers were like, 'Oh, we can't hear them anymore, so there are no longer spotted owls here,'" says Hartman. "But we knew better. I had been surveying using the standard method, driving all night to specific call points, playing recorded calls from ridgetops. Often, I heard nothing at all. But I'd look down at the forest below and I knew there were owls in there, being quiet. We just needed a different way of finding them."

The team decided to try a new owl-spotting technique, this one nose based.

Enter Max, a "detective dog." Hartman trained the blue heeler mix to use his 250 million olfactory cells (more than twenty times the number we possess) for the highly specialized task of detecting the pellets owls eject at roosting or nesting spots. Max, like other detection dogs, was highly motivated by play. "Fetch obsessed" is what they call it. Hartman trained him to sniff out the pellets of both Northern Spotted

Owls and Barred Owls using play as his reward. Smart, spunky, and courageous, Max took to the work like a duck to water.

"Everyone in the field said, 'Don't use dogs to find owls. We already have this protocol, and it works,'" says Hartman. "Only it *wasn't* working."

Hartman and Max participated in a study showing that dogs specially trained in this way could detect the owls better than vocalization surveys and could cover a much bigger area. The probability of detecting Northern Spotted Owls after six vocalization surveys was 59 percent. After three dog searches, it was 87 percent.

"It was definitely a team effort," says Hartman. "I was the second observer on the team, using my experience in the field to guide Max into areas that looked like suitable habitat. But Max was the first observer."

Hartman had been spending so much time in those forests that she had begun to think of them as her backyard. But when she started working with Max, she saw things she had never seen before. "He would lead me to a feather here, a whitewash there," she says. "He had this long swishy tail that would sway, kind of sashay very subtly, when he was on odor. It was wonderful to watch him work, to observe how thorough he was, how he would engage with all the odors and parse them. He was like, 'I'm smelling a lot going on here and I just need to take a moment to catalog it.' Then he would come back and be like, 'This one!' and gently lie down by the pellet."

The day Max found a family of four owls using only his nose, Hartman knew this was an important way of locating these owls. And it was completely noninvasive. "We didn't have to see the owls or elicit hooting from them and attract unwanted attention."

Hartman now runs an organization called Rogue Detection Teams devoted to training shelter dogs for the purpose of surveying for sensitive species around the world (motto: "Conservation unleashed"). Dog detection is what some researchers might call a "fringe method" for finding owls, but under certain conditions, it can be highly effective. "When you're trying to find a really rare animal and there may only be a few of them, or you're trying to cover a really big area in a short amount of time with limited resources and you don't have a lot of people," says Hartman, "a dog team can be really powerful. This is where we lend our noses, so to speak, in the cases where there isn't funding for a big human team."

The Difficult Bird Research Group, a small band of dedicated researchers who study Australia's most endangered species, has recently gone rogue to find endangered Tasmanian Masked Owls. These owls are very rare and deeply secretive. They roost where they are least likely to be disturbed, in cliff-side caves or tree hollows or in the remaining old-growth forest of Tasmania, so they're extremely hard to detect. The researchers have tried broadcasting calls to elicit a response, but with no success. Nicole Gill, a detection dog trainer and field ecologist who collaborates with the Difficult Bird Research Group, has trained a springer spaniel–border collie mix named Zorro to find the masked owl pellets and lead the researchers to the elusive owls. Gill says the "sprollie" will do almost anything for the reward of a squeaky ball. Because other birds, such as Wedge-tailed Eagles, produce pellets similar to those of the masked owls, Gill has trained Zorro to distinguish between the pellets of the two species. "Zorro learned this really quickly," she says, "in a matter of hours, if not minutes. When he hits the scent cone for a pellet, his body language changes. He gets visibly excited. His tail starts to wag faster, and you can see him really focus on finding the odor source. When he locates it, he drops to the ground, and puts his nose by the pellet, and then stares at you expectantly for his ball."

Dog detection may be a marginal method for locating owls. But the notion of one nonhuman animal species working to protect another strikes me as both moving and illuminating. That Max and Zorro can consistently succeed at detecting owls with just their olfactory expertise points to the many ways of knowing in the animal world that go beyond our own.

For finding owls in a vast landscape, the vocal survey and sniffer-dog methods aren't feasible. So how to scope out owls in a giant swath of land?

When a team of conservation biologists was tasked with surveying for Barred Owl and California Spotted Owls throughout the northern Sierra Nevada, "the prospect was overwhelming," says team member Connor Wood. "We looked at the logistics and realized there's no way we can do this with traditional methods." So the team devised a system for "passive acoustic monitoring," planting 200 audio recording devices across thousands of square miles of mountainous terrain to collect owl calls for two years.

It was a massive effort with an urgent goal. Barred Owls are native to eastern North America, but over the past century they've moved across the Great Plains and into the forests of the far West. There, they have effectively colonized the entire range of the Northern Spotted Owl, says Wood, displacing their slightly smaller cousins.

"These two species just can't coexist. They're too similar ecologically, and it's the spotted owl that loses because Barred Owls are bigger, more aggressive, and more flexible in what they eat—just all around competitively superior. There are many factors driving the decline of spotted owls, like habitat loss," he says, "but Barred Owls are an existential threat."

Now Barred Owls are moving south, invading the Sierra Nevada, which is the native range of the California Spotted Owl, a close relative of the Northern Spotted Owl. "And there's really nothing to suggest that things would play out any differently between Barred Owls and California Spotted Owls," says Wood, "so the big question in our minds was: What is the size of the Barred Owl population in the Sierra Nevada, how fast is it growing, and what is its effect on the native spotted owls there?"

Wood and his colleagues gathered 200,000 hours' worth of audio recordings. Then they analyzed them. Wood is nearly tone deaf, which would seem to make the task of analyzing owl calls difficult. But he created custom tools that allowed his team to scan the audio for likely owl vocalizations and then review those manually. It took about 2,000 processing hours per species, he says, plus many hours manually confirming or rejecting the results. He was also able to read the spectrograms, the visual representations of birdcalls and birdsongs. "The dramas that are played out on the audio recordings, I can *see* them. Once you get familiar with spectrograms, you can look at them and sort of *live* these moments that are playing out every night across a whole landscape that collectively represent the reality owls inhabit."

Wood was deeply surprised by what he discovered in the recordings: First, the density of Barred Owls was still fairly low in the Sierra Nevada. And where the bigger owls were present, California Spotted Owls were making *more* territorial calls—the opposite of what had been found in the Pacific Northwest.

"Basically, Barred Owls are really physically aggressive," explains Wood. "Spotted owls quickly learn that if they're vocalizing, this pisses off the Barred Owls, who don't hesitate to swoop in and physically attack them to claim their territory. So where Barred Owl densities are really high, as they are in the Pacific Northwest, Northern Spotted Owls have gotten really quiet"—as we've seen. But Wood and his team found the opposite in the Sierra Nevada, where Barred Owl densities are still low. "This suggests to me that the California Spotted

Owls are still putting up a fight," he says. "They're still like, 'No, no, this is *my* territory, and I'm going to defend it.' Which is pretty cool and kind of heartwarming in a way." It also has important implications for conservation efforts, including the strategy of removing Barred Owls in the Sierra Nevada to support the spotted owl populations.

"What gets me so excited," says Wood, "is that all of this information is there—at some level, the truth about these owls' lives—it's all present in the audio data. It's just a question of accessing it by asking the right questions, and then we can unlock so much about these birds."

Even passive acoustic monitoring won't work in the most remote places on the planet—like the old-growth forests of northeast Asia and Russia, home to the Blakiston's Fish Owl. The terrain is too difficult and hard to access. So scientists are turning to other high-tech monitoring strategies—satellite imagery and drones.

Blakiston's Fish Owl is the world's biggest owl, "the size of a fire hydrant with a six-foot wingspan," says Jonathan Slaght, author of a brilliant book about the birds, *Owls of the Eastern Ice*, and one of the few people in the world who study fish owls. Imagine a fire hydrant–size bird up in a tree. The birds are also comical, says Slaght, slightly wacky looking with droopy, tousled ear tufts and a habit of walking along the ground hunched over. But they're terribly difficult to track down. Both rare and widely scattered, they range from the eastern woodlands of Hokkaido, Japan, to the Primorye territory in the snow-covered Russian taiga and north into the subarctic.

In the Sikhote-Alin area of northern Primorye, one of Slaght's colleagues, Rada Surmach, and her collaborators have worked out how to locate the owls using drones in territory that otherwise might be impossible to access because of rough terrain and deep snow. "The area is very wild, very pristine," says Rada, a PhD candidate and researcher at the Federal Scientific Center of the East Asia Terrestrial Biodiversity

Blakiston's Fish Owl

in Vladivostok, who often works with her father, fish owl expert Sergei Surmach. "In the far north, where the fish owl lives, the temperatures are harsh, −20 or even −25 Celsius in the really cold months," she says. "It's mountainous. There are no paths, and it's physically very demanding. You have to walk across rivers while balancing on a log, and below you is freezing cold water."

To narrow the search for the fish owls, Rada studies satellite images that reveal areas of open water in winter—likely habitat for the owls. "The fish owls nest along rivers where the water doesn't freeze," she explains. "First we locate the unfrozen areas from the satellite images, and then we go there with the drone to confirm that it's really not frozen in the coldest part of winter." If you're on foot, "you can walk along the river, maybe three hundred meters, even if it's frozen, but the drone can fly for five kilometers." There are glitches, however. The rivers run through hills that can block the drone signal. "It's all mountains, mountains, mountains," she says. "You can fly the drone higher, but then you can't see as well—so we're working on creating signal-enhancing devices."

The drone is also useful for checking nests. The fish owls nest in the holes at the top of broken trees, and they tend to use the same tree again and again, says Rada. But checking to see whether a nest is occupied is challenging. "We can take a huge stick and bang it on the tree and hope a nesting bird will flush. But if she's sitting on eggs, she won't move. The alternative is climbing up to the nest, which is often thirty feet up or higher."

A drone eliminates this laborious and dangerous work. "It's just *vwoop*, and it's up there. You can check the nest remotely, and if there's nothing, you can just move on." It also eliminates the problem of leaving a scent trail to the nest that a mammalian predator like a bear could follow.

Next up, Rada and her colleagues are working on a joint project with the Wildlife Conservation Society and the Russian Academy of Sciences to use the new technology to look for fish owls on remote Sakhalin, an island between Japan and Kamchatka. The owls used to be present on the island, but they seem to have vanished. There are a few spots that once reliably had fish owls and no longer do, but no one has ever checked thoroughly. So the team plans to use satellite imagery and drones to remedy that, with the ultimate goal of reintroducing the owl.

Once scientists manage to find their owls, capturing them for study can be a magician's trick. Owls are wary, and researchers often have to be highly inventive.

To catch big, elusive Australian owls, such as Powerful Owls and Sooty Owls, ornithologist Rod Kavanagh and his colleagues devised a complicated trapping system high up in the trees. Using a pulley system, the team rigged the nets twenty or twenty-five feet up. Then they set up two speakers, one on either side of the net, and played the owls calls from alternating speakers so that in one instant the call would be coming from one side of the net and then from the other side. "The owl would fly to and fro, and with luck, it might hit the net," says owl researcher Steve Debus, who used the technique on Barking Owls. "As soon as it did, you dropped the net down, catching the bird. It all took a fair bit of trial and error to get the system to work."

Biologist Dave Oleyar encounters similar challenges using mist nets to catch small forest owls in the American Southwest. To get the mist nets up near the canopy, he and his team use painter poles that extend to twenty-four feet. "We have team members play the audio recordings while they move back and forth around our nets to create the image of an intruding owl moving around, with the hope of luring the real owls into the net," he says. "It's a magical experience seeing or hearing the owls respond to your movements with the audio lure. We essentially fish for owls in the dark with sound, and the owls go over our nets and around our nets and under our nets. But every once in a while, we get lucky, and it works."

Some researchers try grabbing owls using actual fishing nets mounted on long poles, or bow nets—spring-loaded nets designed to catch birds on the ground. Some use a bal-chatri trap, literally "boy's umbrella," an adaptation of an ancient technique devised by East Indian falconers. In the old days, it involved a cage made of split bamboo containing live lure prey, a mouse or a bird, and affixed with horsehair nooses to entangle the birds' feet. These days, the traps are made of chicken wire or hardware cloth for the cage and nylon nooses, but the concept is the same.

"As bird people know, certain traps work for certain species; other traps don't," says Jonathan Slaght. In the case of the Blakiston's Fish

Owl, Slaght faced the challenge of catching an owl that had never before been caught for science. "We really didn't know how to do it," he says. Slaght tried out various ideas suggested by raptor specialists, but nothing worked. "It was a low point in my life," he recalls. "We're out in the woods, living in a tent. It's getting to −30 degrees Celsius at times. The owls are there. They're vocalizing. We're setting the traps. But they're not coming in. They're not getting caught." This went on for weeks until Slaght came up with the idea to use a prey enclosure set in the river where the owls hunt for fish, essentially a mesh box filled with fish sitting in the water. "Once the owls discovered the prey enclosure, they hung around and guarded it jealously," he says. Then it was a simple matter of setting a "noose carpet" on the shores of the river to capture them as they made their way down to the prey enclosure. Talk about creative.

Sometimes owls outwit even the most experienced and ingenious trappers. But here's the thing: *why* individual owls are sometimes difficult to catch may offer up important information, signifying something vital about the nature of the species.

DIGGING INTO BURROWING OWLS

It's November, going on summer in the southern hemisphere—a world away from snowy northern Primorye. It has been a long day with no owl captures. A small team of us has set ten traps in the burrows of Burrowing Owls at a half-dozen sites in the vacant lots around Jardim Oriental, a suburb of the leafy city of Maringá in southern Brazil.

I'm here to learn how owls are studied in the field with one of the world's foremost experts on owls, David Johnson. The sun is hot. The soil is red. We're all covered in a sheen of red-bedecked sweat from digging in the dirt to wedge special traps firmly in the mouths of the burrows. This project, aimed at exploring the family tree of Burrowing Owls, is part of Johnson's larger effort to understand everything he can about these charismatic little owls.

Burrowing Owls can hardly be mistaken for any other species. They're cartoonish, like a two-legged head, with long, almost stilt-like legs, a short tail, and a compact, expressive head that's often cocked or turned in what looks like curiosity or an effort to get a different perspective. They're funny to watch, natural clowns, with a habit of bobbing up and down when agitated, as James Bond wrote in his book on the birds of the West Indies. (Yes, that Bond. The model for Ian Fleming's spy was an ornithologist and published a guide to Caribbean birds in 1936.) Contrary to common belief, they're not day creatures, but nocturnal and crepuscular hunters. They eat insects, small mammals, reptiles, amphibians, and even small birds. In the day, they have a kind of slow, undulating flight, but at night, they're the nocturnal equivalent of kestrels, quick, agile killers.

The owl's Latin name, *Athene cunicularia*, means "burrower" or "miner." Typically, Burrowing Owls use ready-made burrows dug by prairie dogs, woodchucks, skunks, badgers, armadillos—really any fossorial mammal—and even tortoises, which spares them the trouble of excavating their own holes. They'll also use human-made structures, such as cement culverts, debris piles, or openings beneath asphalt pavement. But here in southern Brazil, many dig their own burrows wherever the ground is inviting, and their reddened feathers show it. When museum experts were examining Burrowing Owl skins from this region, they remarked on their coloring and assumed the birds were red morphs, with feather pigmentation that differed from their northern relatives. In fact, they were just dirty. My shoes are stained a rusty red, along with my fingers and the pages of my notebook.

Burrowing Owls

Burrowing Owls like wide-open grassland plains and prairies but will settle for pocket prairies and mini-grasslands in more urban environments. In this region of Brazil, with its fertile red soil, their natural environment has been transformed into pasture and farmland, forcing the owls to seek a less desirable breeding habitat. In this suburb, they're nesting on a steep grassy embankment behind a university building, beneath the small causeway to an elementary school crawling with children, and in several vacant lots in Jardim Oriental, despite the constant chaos of dogs, motorcycles, loud music, weed cutters, and a steady stream of pedestrians and cars.

In other words, they're ubiquitous. But at the moment, they seem impossible to catch—wary and defiant of the usual strategies for trapping them. They watch us from a safe distance, perched on a fencepost, a ROÇADA sign (advertising a service to clear vacant land), or a little eminence in the corner of the lot, gazing at us with a suspicious air. Later, much later, in the balmy night, when we finally pull the traps in defeat, they'll leave their watch posts and duck into their trapless burrows, as if to thumb their beaks at us.

Southeastern Brazil is one of the world's owl hot spots, with seventeen of the country's twenty-six species. When I ask Johnson why there's such a diversity of owls in this region and other global hot spots—southern Asia, a swath of Arizona and Mexico, sub-Saharan Africa, China—he tells me it's a synergy of two things.

"These are some of the places that have been the most environmentally stable for the longest period of time, millions of years," he says. "They're also geographically varied, with different habitats." They're mostly in tropical regions that never experienced glaciation. "Glaciation kind of wipes things out and resets the whole landscape, and it takes a long time to get any kind of diversification. These are top predators and, inherently, there isn't much room at the top. The only way you get more diversity of species is through diversity of available niches. So it's these two conditions, stability of climate and landscape and variety of topography, that have allowed owls to diversify in these regions."

Here in southeastern Brazil, habitats vary widely from sandy coastal plains and grasslands to the Atlantic Forest (which bursts with diversity and so many distinct habitats that it harbors around a thousand bird species, many of them endemic to Brazil). Here there are Stygian Owls and Spectacled Owls, Mottled Owls and Striped Owls, Tawny-browed Owls and Buff-fronted Owls, as well as three species of screech owls—among them, Long-tufted Screech Owls, which are found nowhere else in the world—and also Burrowing Owls, found nearly everywhere else, at least in the New World.

Burrowing Owls range through twenty-four countries in the Americas, from Canada to Chile, Barbados to Mexico. There are twenty-three subspecies, including two that have gone extinct. Johnson's project here in Brazil and across the Americas aims to query the current taxonomy. Are Burrowing Owls in fact a single species, or should they be cleaved into two or more species? Do some subspecies that have been dumped into the same bucket just because they look the same (like barn owls once were) deserve to be in a species category all their own? Is a Burrowing Owl in Canada the same as its cousin in southern Brazil? Among the subspecies of Burrowing Owls in the Americas, is there a new, true species hiding?

To sort out these questions, Johnson and his team are gathering information on the range and distribution of all twenty-three subspecies and "sampling" them, capturing adult owls and taking morphological measurements and blood samples for DNA studies on a minimum of thirty adults of each live subspecies: fifteen males, fifteen females. For the extinct species, he's using museum specimens, measuring the skins and extracting DNA from the toe pads. In the live owls, he's also recording the vocalizations of at least ten different males of each subspecies. "One of the ways that you can differentiate species is by their calls," he says. Two species may look alike but sound very different. Altogether he anticipates gathering useful data from a total of 2,000 live owls and museum study skins.

The questions Johnson seeks to answer may seem esoteric, but they go to the root of some of the biggest issues in owl biology. There are similar questions for all kinds of owls around the world, screech owls, pygmy owls, scops owls, hawk-owls, boobooks, and barn owls. It can be hard to draw distinctions between birds that differ only slightly. How many owl species *do* we have in the world? How do we tell a species from a subspecies? How do species of owls diverge? And what can owls tell us about how we might define a species, such a fundamental unit of life? The taxonomy of owls has been a problem for centuries. Now, with new, modern tools, including powerful cutting-edge genome-sequencing technology, Johnson hopes he can parse the mysteries of the Burrowing Owl family tree. But to do so, we'll have to catch dozens of owls here in Maringá, and they're not making the job easy.

David Johnson has more experience than anyone on the planet trapping Burrowing Owls. He estimates the number he's captured at around 6,000 worldwide. Much of what we know about these owls, we know from his research. For fifteen years he has led a long-term study of Burrowing Owls in the Pacific Northwest, which is unearthing strange and wonderful new findings about how these owls pick their mates, whether they inherit their vocalizations (passing them down from great-great-grandparent to great-great-grandchild), how they choose their burrows and decorate them, what they eat, where they go when they migrate—all in all, some sixteen different aspects of owl ecology, biology, and conservation. Over the course of the study, he has tracked the outcomes of more than 660 nesting attempts and trapped and banded more than 2,600 owls at this site alone.

Like many owl researchers, Johnson fell in love with owls at a young age. When he was eleven and camping alone along the Blue Earth River in southern Minnesota, an Eastern Screech Owl landed on his canvas pup tent. "It called for a good twenty minutes," he recalls. "I was sitting inside, and the owl was literally just inches from my face." He could see it in silhouette and watched its body vibrate through the tent as it trilled. "I didn't know what species it was at the time," he says, "but it certainly made an impression. And then I realized . . . that owl could sit in the trees anywhere around there, but it was sitting on my tent. So I took it as a special message."

Johnson likes to say that there are two important days in life, the day you were born and the day you find out why. Now he runs the Global Owl Project, a consortium of more than 450 researchers and conservationists in sixty-six countries working on the science and protection of owls. He travels all over the world, collaborating with colleagues, giving talks, and mentoring young researchers and PhD students, which is one reason he's here in Brazil. He had arrived in Maringá a few days earlier, after a good twenty-four hours of travel from the United States, with an oversize box full of trapping equipment and a head start locating the owl burrows—thanks to the efforts of Priscilla Esclarski, the Brazilian leader of the research team and one of Johnson's protégés. Esclarski and another team member, Gabriela Mendes, had publicized the study and sent out inquiries to the public asking whether people had seen nesting Burrowing Owls. They got a flood of responses, followed up on the leads, and pinpointed several burrows. Johnson is here to advise them on strategies for catching the birds and collecting data on them.

Both Esclarski and Mendes came to this field through their long interest in corujas, the Portuguese word for owls, and in the myths and beliefs some Brazilians carry about them. Esclarski says she was deeply interested in birds as a child. She identified with owls, birds of the night, and wanted more information. "There was no event as remarkable as David's in my life," she says, "but I always knew that I was born for them (or they chose me? I don't know!)." In one of her first research projects, Esclarski explored folkloric elements, legends, songs, and Indigenous traditions about owls in different cultures within Brazil, including quilombolas, Afro-Brazilian residents of the settlements that were first established by escaped enslaved people in Brazil. Later, she joined David Johnson's Global Owl Project and interviewed people of all ages and in all regions of the country about their beliefs about owls. Mendes joined Esclarski on the owl project for her graduate studies. Their work took them into schools in farming communities, where "the children told us their parents still thought owls would do something bad to families," says Mendes. "They threw stones at them and even killed them. They had no access to information. They would hear the owls and start to imagine things."

Now Mendes is studying the effect of urban environments on Burrowing Owls, and Esclarski is heading up the investigation of Brazilian subspecies for this project. Both want to understand the puzzles these owls present and to educate the public about their nature. Thanks to the prep work, the owls' nesting sites have been fairly easy to find. Once we locate the burrows, we look for telltale clues of whether they're active—pellets and whitewash from the owl feces, and trampling around the site, which suggests that young are present. Also, scattered feathers or bits of plastic, cloth, Styrofoam, sponge, and broken glass that males have used to decorate their burrows here in this urban environment, and sometimes crippled insects they have brought in to feed the family.

It's the job of a male Burrowing Owl to deliver food to the burrow to feed the female. She likes fresh food, so the male doesn't kill the insects, just disables them, says Johnson. (Rodents he won't leave alive because they'll escape.) He delivers spiders and moths and other insects and leaves them within a foot or so of the burrow mouth. "When you walk up to the nest site at night, you'll see the eye shine from a lot of insects and spiders," says Johnson. At one nest, he found thirty-two wolf spiders. ("That guy specialized in wolf spiders.") The insects and spiders are still alive, but they can't get away. "How do the owls know to do that?" he wonders aloud. "The male goes off, catches something, cripples it, brings it back, drops it off, and then goes to get another one. That's impressive."

Once we find the burrows, we set the traps, which Johnson designed himself. They're black rectangular wire mesh boxes eighteen inches long and six inches high, with swinging trapdoors on both ends. Outside the trap, he positions a small audio recording device about twenty inches from the burrow entrance to record the vocalizations of the owls. And just inside the trap, he sets up an MP3 player that softly plays the recorded territorial call of a young male Burrowing Owl from Oregon.

This is Johnson's secret weapon for trapping these owls. How he discovered it is a remarkable tale of owl sleuthing and attempting to fathom what's going on in an owl's mind.

The story begins at a US Army weapons depot, an unlikely spot for an owl haven—and one with a peculiar backstory that shows just how tightly communities of creatures are knit together, including owls. Pull at a single thread and the whole fabric may unravel.

In 2007, the army called Johnson to ask if he could help them with a problem at the Umatilla Army Depot, an old chemical weapons storage facility on an open stretch of 17,000 acres near the border of Washington and Oregon. "A significant percentage of the nation's stock of lethal chemical weapons had been stored at the depot, including sarin gas and mustard agent," says Johnson, cached in 200 or so concrete igloo-like bunkers. After the United States signed a treaty with 192 other countries in 1997, agreeing to dispose of chemical weapons, the army built an incinerator and ran it 24-7 for ten years to destroy more than 3,700 tons of the weapons. Then in 2012 it decommissioned the depot. All that remained were the empty concrete igloos, giving the place the feel of an abandoned lunar colony.

Even before the weapons were destroyed, things had gotten interesting at the depot, ecologically speaking. In 1969, the Oregon Department of Fish and Wildlife had brought in fourteen Pronghorn, a single dominant male and a harem of females, to populate the place. It got more than it bargained for. The Pronghorn population quickly grew to 200 or 250 head until it suddenly crashed. The reason? "The animals were confined by a perimeter fence, so they became both overcrowded and inbred," says Johnson. "Because of inbreeding depression, they lost their genetic heterozygosity, and the herd began falling apart."

But the land managers didn't see it that way. They blamed coyotes for the population crash. "They said, 'Oh, no, the coyotes are getting the fawns, so we have to implement a coyote control program,'" says Johnson. In trapping the coyotes, they also trapped as bycatch all the badgers living at the depot. "They zeroed out the badgers, eliminating the badger burrows, and the lack of burrows cascaded down and caused the Burrowing Owl numbers to crash because that's where they were nesting." This was problematic because the owls kept in check the rodent population at the depot.

That's when Johnson was summoned. "They called me up to say, 'David, we're losing our Burrowing Owls—what can we do?'"

Johnson had a solution: In 2008, he and his team installed nine artificial burrows designed to mimic natural burrows, but also to be accessible to researchers. For the nesting chamber, he used fifty-five-gallon barrels cut in half, with a lid that could lift off for a scientist's easy access to the nest. For the tunnels leading to the nest chamber, he attached six-inch-wide PVC tubing ten feet long, which is about the length of a natural burrow tunnel. The entrances were armored against dogs and coyotes with rocks. Eventually, Johnson installed ninety-six of the human-made burrows.

The owls instantly took to their new homes. Before the artificial burrows were installed, there had been only three or four pairs of nesting owls. The year after the installation, there were nine pairs; the following year, thirty-two pairs; then, sixty-one—until, in 2021, the population peaked at sixty-five pairs, with some nests producing as many as ten chicks. A truly impressive rate of increase. "Man, it was a packed house," says Johnson. "All in all, we had three hundred fifty-eight young that year, all healthy, happy, and robust."

Johnson was delighted. From early on, he understood he had a ready-made opportunity for a long-term study of a sizable population of Burrowing Owls. "I realized we could band these owls, put geolocator tags on them, and learn about their mate selection, diet, dispersal, migration patterns, choice of burrow designs, all kinds of stuff," he says. He could get to know the birds individually, exploring changes in the owl population over generations.

That meant catching every owl.

At the time Johnson began the Umatilla study, the main technique for capturing Burrowing Owls was a walk-in trap made of stainless steel, wire mesh, and burlap. "And people were telling me that using this strategy I should be able to catch 70 percent of females and 30 percent of the males," he says. "And I thought, 'Okay, that's good. But I don't want to catch *some* owls. I want to catch *every* owl. I want to catch every owl in the study area *every year*. And every place I go in my Global Owl Project work, I want to train people to catch every owl.'"

Johnson modified the trap, painting it black with a lighter mesh door so it looked more like a tunnel extension and was less visible to the owls. He also used a bow net with a little solar-powered grasshopper or cockroach on top of the bait cage to catch the owls in the daytime. "They have little wire legs and they kind of buzz as they wiggle around. I put them on top of the cage and they just *bzzzzz*, which works great when the owls can see it." And at night, he placed a fake or dead mouse inside the cage, along with a little MP3 player that broadcast a mouse

Burrowing Owl chicks in an artificial burrow at the Umatilla Army Depot

sound. Between the two traps, he could catch the female owls and about 95 percent of the males. "But I wouldn't catch them *all*," he says. "The old-timers, the really good hunters, would fly over my bow net and say, 'Thanks, David, but I can do better myself. I can catch kangaroo rats. I don't want to fuss with your little mouse. I know it's a game and I'm not playing.'"

How to capture those last few canny old birds?

The answer came in a flash of insight from an unexpected place: the owls' migratory habits. To try to understand migration in these owls, Johnson had put geolocator tags on several males and females. This was a feat in itself.

The geotags, tiny data loggers and transmitters, are a great way to collect an owl's location several times a day, providing abundant data about its activity, where and when it's resting, moving, hunting, or migrating. But the tags are costly, and while some of these devices transmit the data to a satellite or receiver station, like a cell phone tower, others require recapturing the bird to download all the information—which can be a gamble. Moreover, the owls are sometimes ingenious at finding ways to shed their expensive baggage.

Johnson experimented with methods for securing a transmitter to the back of a Burrowing Owl with a little backpack to make it as lightweight and uncumbersome as possible. "People had been using forty-pound test nylon-coated stainless steel cable on Peregrine Falcons," he says. "So I thought, well, owls are tougher, so let's double that." He put on eighty-pound test cable. "Half of the owls chewed through it," he says, "just chewed through stainless steel cable. And I thought, 'Wow, that is an expensive lesson.'" He showed me a video of the escape action. "Here's the male helping the female bite through her cable," he narrates. "And here she is dropping the cable off her leg. They're doing a tag team thing—give me a break!" This sort of cooperative "rescue" has been documented in Australian Magpies and is considered a kind of altruistic behavior, helping another individual without getting an immediate, tangible reward. Remarkable conduct, but it can vex a researcher. Eventually Johnson found a material so strong it's used for suturing in medical applications. Since then, not a single bird has shed its backpack.

All this effort was worth it. The geotagging revealed that only female Burrowing Owls in the region migrate south in winter, heading to a safe place with abundant food so they can put on weight and get into shape to lay their eggs. Males, on the other hand, stay around or head north to a nearby area.

They head *north* in the winter?

At first Johnson thought there was something wrong with his data. Why would males choose to winter in a frigid area where food is scarce? He realized there could be only one reason: "Because they want to be first back to their burrows! Their strategy is, 'If I can survive in the north, I get to be first back. And if I'm first back, I get dibs on the best burrow and the best territory.' For a male, real estate is worth putting your life on the line."

Once he understood how important breeding burrows are to male owls, Johnson realized that he could lure even the wiliest bird into the trap by making him think his coveted burrow was being invaded by a rival. For the audio lure, Johnson uses the call of an underweight eleven-month-old male, "a punk," he says. "I don't want the call to sound like a dominant competitor, just a pesky intruder, somebody that's easy to shake out of there." When the intruder doesn't show himself, the owner usually hoots his own territorial vocalizations from near the burrow, which are recorded by the audio recorders. If his calls don't draw out the phantom intruder, the male will go in after him. "And boom! I catch him," says Johnson.

"The birds that take the longest to enter the trap are actually the shy young males because they don't have the chutzpah to go after this interloper," says Johnson. "They just hang out outside the burrow and talk about it. They strut around and they call. I get awesome recordings from these guys. If I leave for an hour and then I check on the recorder, I'll have five or six hundred territorial calls. Then as the guys get older, five-, six-, seven-, eight-year-olds, they don't have time to talk about it. They'll give me six calls and then they'll trash my equipment. They'll kick it around. They'll fuss, tip it over, jump on it, or charge the DeadCat [the wind muff covering the microphone] and grab it, whisk it away. They know the routine and they're mad. The recorder will pick up the swoosh of raining sand as they kick up a storm. Then they go inside their burrows to see who's there, and I've got them."

Johnson's success rate capturing Burrowing Owls in Oregon with this method is close to 100 percent. "It's all about getting inside their minds, thinking about what motivates them," he says. "Once you do that, they're yours."

These days, it takes Johnson an average of twenty minutes to nab a male. His record is forty-five seconds. But that's at the depot in Oregon. Here in Maringá, 6,700 miles to the southeast, the nest burrows set with traps and MP3 players broadcasting those territorial calls of a wimpy male Burrowing Owl from the Pacific Northwest remain empty.

Why aren't these birds drawn in by the usual trickery, the call of the male recorded up north? They should be diving for their burrows to defend them. That is, if they're the same species. But are they?

We wait for hours in the car for the owls here to fall for the ruse. The birds are going everywhere but into the traps—loitering around the entrance to their burrows, perching on nearby bushes and fenceposts, making short flights at low elevations from one lookout post to another, or just standing bolt upright in the middle of a lot and observing us closely. We watch a female dig under the trap in her burrow to get at the young inside. We get out and walk around the birds on the red, pocked soil, avoiding eye contact to try to flush them into their burrows. "They watch your eyes," says Johnson. "If you put sunglasses on, they lose interest, just like that. Eyes and eye contact are fundamental to how they measure danger."

Then finally! A pair of bright yellow eyes glowing in the dark of one of the traps, like welcome little beacons. At last we have snared a bird, a slim start to the gathering of clues. This trap had been set in a burrow in a vacant lot on a noisy road opposite a building where loud, thumping music blared. It's hard to imagine how the owls could tolerate all the human activity. Both male and female did seem wary, standing guard near the burrow entrance. When Johnson approached, the female ducked into the tunnel with its trap. Now she was caught.

Burrowing Owls in a trap

I'm deployed to check the trap with a flashlight. When I crouch down, I see those eyes burning bright in the dark and wave excitedly to Johnson. He hurries over and reaches into the trap with a hand, grasping the elfin owl firmly, turns her over facedown, and draws her out. Then he cradles her in his arms, gently, tenderly, like a baby. She squeals and struggles and fixes Johnson with a furious glare.

This feisty female is Burrowing Owl number one for the Brazilian arm of this study. Johnson knows the owl is female because she's bigger than her male counterpart would be, and darker. During the nesting season, the feathers of males bleach out in the sun as they stand guard over the nest.

The processing station is set up behind Esclarski's car on plastic bins and cardboard boxes, and the team members—Johnson, Esclarski,

Mendes, and two additional helpers, Thaís Rafaelli and Vinicius Bonassoli—work quickly, photographing the owl, measuring her hallux talons with calipers and assessing her wing length, area, and span while she clacks loudly, and then tracing the outline of both wings, one at a time, on a white sheet of paper, as if she's flying.

All in all, they collect twenty-five different morphometric measurements of body and wings. It's delicate work requiring gloveless hands, and Johnson's fingers are stuck again and again by the owl's sharp little talons, designed to take prey in the air. He draws her up to examine her closely. "Whew," he says, "you have insect breath." Her brood patch, the bare skin that heats her eggs, is huge, up to her chin, but her feathers are growing back, suggesting that she is either far along with her youngsters or that her nest failed. Esclarski puts the owl in a cloth sack to weigh her. She's heavier than most Oregonian Burrowing Owls. It takes three team members to draw a blood sample with a syringe from a tiny vein beneath her wing—just four or five drops—and siphon it into hematocrit tubes. With the tubes, they draw spirals of blood on a strip of sensitive paper that will be used to extract her DNA.

The full processing takes only about fifteen to thirty minutes. When they're done, Johnson carries the owl back to her nest, reaches his arm all the way through the back of the trap, and deposits her in the burrow. Then he latches the back door of the trap so she won't get caught again. He leaves the front door unlatched to try to capture the male, who is eyeing us from atop a stop sign.

One bird does not a study make. In the following days, Johnson shifts his strategy. He has recorded the owls' territorial calls and listened closely to them. It's clear they sound nothing like the calls of the owls in the Pacific Northwest. The Oregonian owls have a monotone two-note call, *coo-coo.* Here in Maringá, the calls have quick up-and-down rhythms, more like *coo-kyeeia,* and more time between vocalizations.

If the calls are different enough, the males here may not recognize them, says Johnson. The territorial call on the MP3 player "probably just sounds like some strange thing in their burrows. They recognize it as a bird, yes, but maybe a sparrow or some other species." In other words, the Burrowing Owls here don't speak the same language as their relatives in western Oregon. "Their last conversation was maybe four million years ago between their last common ancestors. That's just a guess. But the amazing part is that we're getting to witness this. If it plays out in the analysis, we'll know how separate these owls are from their North American cousins in space, time, and evolution."

With a new recorded territorial call of a Maringá owl that Johnson has loaded onto his MP3 player, the team members start capturing owls right and left, some with their meals en route to the burrow, a mouse or a gecko minus the tail. In a matter of a few days, they have an adequate sampling of this local population and can move on to new areas of Brazil.

From Maringá, the researchers travel the length of the country, heading out in the dark to capture the owls at sites ranging from the long narrow beaches of Lagoa de Peixe National Park at the southeastern-most tip of Brazil (where the owls decorate their burrows with natural detritus, the jaws of small mammals, the flippers of small turtles, and mollusk shells) all the way to Boa Vista in the far northern Amazon Basin, a distance of almost 2,500 miles. They learn that the owls in Maringá (the *grallaria* subspecies) are 20 percent bigger than those in western Oregon, heavier, tougher, stronger, and more robust, which runs counter to the prevailing idea that in warmer climates birds are smaller. They discover that the Burrowing Owls in the south don't respond to the territorial calls of Pacific Northwest males, while those farther north, in Boa Vista, do. In Boa Vista, they find the smallest subspecies of Burrowing Owls, known as *minor,* close to an area with a larger subspecies (probably *grallaria*). In fact, says Esclarski, "we found a pair in which the male was possibly a *minor* and the female a *grallaria*. But only genetics will tell us who is who. And that's exciting."

All over the Americas, researchers are studying the owls in this way for Johnson's project—in Peru, Ecuador, Colombia, Venezuela, Mexico, Aruba, the Bahamas—all with the same equipment and methodology, all with the same goal.

The DNA samples the teams gather from live captured owls, along with those from the extinct subspecies, will be processed at the San Diego Zoo Institute for Conservation Research. The institute will sequence the DNA of all the different subspecies to determine their ancestry, working backward to calculate the time of the owls' most recent common ancestor.

To complement the genetic analysis, Johnson will compare all the morphological measurements, as well as the vocal recordings of all the owls' territorial calls, and analyze them for vocal variation using special software that can differentiate the pitch and tone of the calls, their frequency and harmonics, the time between notes, the time between calls, how many calls in a sequence, and other features.

"The question is," says Johnson, "when you compare all these genetic, morphological, and vocalization data, are there significant enough differences in all these facets to recognize a distinct species? Are the feathering and distribution different enough? Has evolutionary divergence of vocal patterns produced a significant enough wedge between populations that they're no longer interbreeding? Does this change in vocalizations reflect genetic divergence?"

To address these questions in any bird family takes large sample sizes, resources, a lot of time, and—in this case—cooperation from two dozen countries. David Johnson is excited. "All of a sudden, answers to the puzzle of Burrowing Owls in the Americas are within reach. It's because of technology. And partners." To study an owl takes a village—or in this instance, a world of villages.

In bringing to bear the new tools of owling, Johnson's project may grow the number of owl species by at least one, perhaps more. No conclusive results yet, but some tantalizing clues, especially as revealed in the differences in the Burrowing Owls' territorial calls. In listening to the variations between the calls of different subspecies, it's possible we're eavesdropping on evolutionary change, witnessing something as revealing as the beaks of Darwin's Galápagos finches in generating new species.

And those territorial calls are just the tip of the iceberg when it comes to owl communication. Owl calls may sound simple to us, but what do they convey to an owl? What does an owl hear in those few seconds of carefully shaped sounds?

Far more than we ever imagined.

Four

Who Gives a Hoot

OWL TALK

Alice was acting distinctly owly, even by Great Horned Owl standards, peevish, irritable, grumpy. And at the moment, she was having none of Karla Bloem's efforts to appease her.

"She was hooting at me, and I was trying to hoot back to her," says Bloem, "and she walked over to the edge of her perch and smacked me hard on the head with her bill. And I'm like, 'I don't know what I'm supposed to do.' And she was getting visibly upset because I wasn't responding properly."

Alice is what's known as a human-imprinted owl. When she was just three weeks old, she tumbled from her nest at the top of a pine tree in Antigo, Wisconsin, and broke her elbow badly. She was taken to a raptor rehabilitation center before she could fully see and was cared for by people there, which caused her to imprint on humans rather than other owls. Now she's not fearful of people and is psychologically inclined toward relationships with them. This, together with her significant injury, meant she was incapable of surviving on her own. "The

Alice the Great Horned Owl hooting

rehabilitation folks knew from the get-go that she would never be able to live in the wild," says Bloem. "An owl like this has to get a job or she will be euthanized."

Alice did get a job, working with Bloem, executive director of the International Owl Center, a nature center in Minnesota that "seeks to make the world a better place for owls through education and research." Alice's role would be as a kind of educational ambassador to teach people about owls. But first Bloem needed to understand what Alice was saying. This owl might have been raised by humans, but she still had all her normal owlish instincts, including the drive to communicate. She just applied them to people.

"She was making all these sounds toward me and expecting me to respond as a male Great Horned Owl would," says Bloem. "She was getting mad because I wasn't doing it right. In self-defense, I went to the scientific literature to find out, well, what the heck do her hoots mean? And I found out that nobody had ever studied Great Horned Owl vocalizations, which seems weird. I mean, it's a common species all across North America! Why were they not studied?"

It was the spark for a nearly two-decades-long exploration of the calls of Great Horned Owls, their purpose, variety, and significance in the wider world of owl communication. Understanding owl vocalizations is crucial to understanding nearly everything about them, says Bloem, their identity, habits and attitudes, intent, territoriality and preferred habitat, and their relationships with mates, family, allies, and rivals. "Owls basically see their world through their ears. They're mostly active at night or at dawn or dusk, so vocalizations are essential to their communication. They don't just hoot for the hell of it. They vocalize for a reason, and they convey meaning in their calls."

The hoot of an owl is one of the few birdcalls most people know. But a hoot is not just a hoot. There are greeting hoots and territorial hoots and emphatic hoots. And owls don't just hoot. They shriek, yap, chitter, squeal, squawk, warble, and wail plaintively, most often in courtship songs—love songs made of odd and uncouth sounds generally unappreciated except by the ears for which they are intended. Some owls sing with the full power of their lungs; others coo softly. Some chirrup like a cricket. Some chuckle or roar with maniacal laughter. In breeding season, the male Mottled Wood Owl of India utters a blasting, shivery "laughing" call, *chuhuawaarrrr.* The Sooty Owl is known for its strange sliding whistle, like a dropping bomb.

I once stood on a friend's porch in Italy and listened in the dark to the calls of four pairs of Tawny Owls from the deep woods across the way. Vibrant, quavering two-note hoots, slipping in pitch, followed by the sharp chirrups of females—first over here, then from a distance in one direction, then from the other direction. The belligerent little owls were patrolling the boundaries of their territories, animating the dark heart of the forest.

The call of the barn owl has variously been described as a harsh scream, a nasal snore, a series of notes, *click, click, click, click, click,* like a katydid, and a raspy hiss that sounds like a fan belt going out on your car. The Barking Owl I heard in the Pilliga Forest in Australia not only utters a canine *wok, wok* or *wook, wook*, but also growls, quietly at first, then more emphatically. It may utter a soft sneeze followed by an abrupt scream and a tremulous, peevish hoot. The toots, trills, and chitters of the Northern Pygmy Owl match its pint-size body. But the deep, husky voice of the tiny Flammulated Owl belies its diminutive stature. Here's another way owls break rules. In general, body size for a bird dictates the pitch of its vocalizations. The bigger the bird, the lower its pitch. Smaller birds usually have higher, twittery voices. The tiny Flammulated Owl explodes these formulas. It slows the vibrations of its call by loosening the skin around its throat, creating a low-pitched, husky hoot more suited to a Great Horned Owl, says Brian Linkhart, who studies Flams. "It's a big bird trapped in a small body."

JR, an Eastern Screech Owl that lives at the International Owl Center, has a call that sounds exactly like a ringing phone. And when a real phone rings, he responds with his own ringing call, says Bloem. "It's very hard to answer the phone with a straight face, with JR there in the background ringing away."

Even among these weird and wondrous hoots and toots, the outrageous squalling, snarling, growling, howling, and caterwauling of the Barred Owl deserves special mention. Once, when ornithologist Rob Bierregaard was walking through the suburbs of Charlotte, North Carolina, playing Barred Owl calls to attract the birds, someone asked him whether he had lost his monkey. Later, deep in the woods, he heard the owl's powerful bloodcurdling "woman being murdered shriek" directly above him, and it made him jump three feet.

The zoomusicologist Magnus Robb has a gift for characterizing owl calls. Robb describes himself as a musician who "went zoo." Not only does he make crystalline recordings of owl sounds, but the words he uses to describe them are *brilliant, piercing,* and *evocative.* The soliciting call of the female Long-eared Owl is "shaped like a heavy sigh," he writes, "the *Vvvw* descends and fades out towards the end. If you have ever tried the old game of humming through a comb covered in cigarette paper (or Izal, that crispy and perhaps obsolete toilet paper that scratches your backside), well, the timbre is a lot like that."

I don't know about you, but this makes me want to run to the woods to listen.

It's not easy to describe owl calls, much less to imitate them. But at the International Owl Center, the ability to hoot like an owl is written into the staff job descriptions. It's a sort of litmus test of obsession, but also a useful tool when people come to the center seeking to identify an owl they heard. "We ask, 'Well, what did it sound like?'" says Bloem. "And more often than not, people can't remember. So if you can roll through a bunch of different owl calls, then they can say, 'Oh, yes, it was *that* one!'"

The easiest call to mimic may be the territorial hoot of a male Long-eared Owl. Bloem demonstrates a quiet, subtle, evenly spaced *Hoo. Hoo. Hoo.* "Simplest owl call on the planet," she declares. "Even a four-year-old can do it."

The hardest may be the call of a Brown Fish Owl, which has a voice so deep many people can't reproduce it at all. It's not particularly loud, but its low pitch means it can be heard above the thunderous roar of the streams and rivers where the owl fishes.

The lecturers invited to speak at the International Owl Center are always asked to offer their best hoots. The request frequently brings out either the instructor or the ham. Take Brian Linkhart, a professor of biology at Colorado College: "I'm going to teach you what to do with your mouth to imitate a Flammulated Owl call," he says. This is the assignment he gives his students at the beginning of the field season. He calls it "FO Calling 101." The students will be heading out in a few weeks to survey owl populations, and they'll need to know how to hoot "flam style" to elicit a male owl's call. "So your lips are going to form the word *boop*," instructs Linkhart. "But you're not going to sound the consonants. As a human, the sound *ooo* comes out. Good. But these owls are also ventriloquistic. They have the ability to throw their voices and also make them hollow sounding—a defense against predation. To get that hollow sound, open your throat as far as you can. And instead of having the air come straight through your teeth, as we do when we talk, try to direct that air up to the top of the mouth, to the palate. So the sound is going to reverberate around your mouth a little bit before it comes out."

Phew. That is *not* easy, and it gives one a new appreciation for what it takes to be an owl.

"If you practice that religiously for about two weeks," says Linkhart, "you can actually start to fool a male Flammulated Owl. And that's a head trip, because once you get one of these owls to respond to you? Man, you start to feel like you're growing wings. It's just a really cool thing."

Again, I'm feeling the impulse to run into the woods, ear cocked and lips pursed to *boop*.

Why do owls hoot, anyway? Are their calls a kind of language? If so, what are they saying to one another? To allies, enemies, mates, and family? Are owls that sing together duetting? Or are they dueling with their voices? When does an owlet start to hoot, and how does it know to do so?

To tackle these and other questions, Bloem began studying Alice's vocalizations in 2004. Eventually, she expanded her observations to Great Horned Owls in the wild, as well as to a captive pair at the center. "I wanted to understand the different vocalizations and their behavioral context so I could tell what's going on with these birds, even without ever actually seeing them," she says. "If you know the context associated with each type of vocalization, you can make inferences about behavior. All of a sudden, the world opens up and you can see what's going on with just your ears."

It's a world she did not expect to explore. Born in rural Houston County, Minnesota, Bloem says her childhood passions were focused on hawks and falcons. "I grew up on a farm, and my job was raking hay. And when you're out raking in a hayfield, you scare up little critters, and the hawks come and circle overhead. Red-tailed Hawks would follow me around when I was raking. And that's really what spurred my interest in birds." In college in Iowa, she majored in biology and worked under an ornithologist. After graduating, she became a falconer and flew kestrels. In 1997, when the city of Houston, Minnesota, decided it wanted a nature center, it hired the twenty-five-year-old falconer to build the center from the ground up.

She realized she would need to start with programming. "So I was looking for a bird to use for education," she says. "I had my falconry birds, but they were kestrels—great for falconry, but not for education. They're quick learners, but they're high-strung. Honestly, I wanted a Red-tailed Hawk, not a Great Horned Owl. But it was Alice who was available at the time. And that's literally what started the whole owl thing."

It took a solid year for woman and bird to get to know each other. "I'm used to birds of prey being pretty standoffish," she says. "But with owls, it takes even longer. They don't adapt well to change. And Great Horned Owls really are poster children for the word *owly*. Alice can be downright snarky."

During those early days when Bloem was trying to figure out how to communicate with her new owl, she hooted at Alice while standing close to her perch. That's when Alice would smack her head with her bill. "I was a slow learner," she says. "But finally, I tried leaning forward while hooting, and she looked at me with apparent surprise, like, 'Duh! You finally figured it out!' and never whacked me again. Apparently, bowing while hooting is how it's done."

I should say here that living with a Great Horned Owl—or any owl, for that matter—is not easy. Bloem's freezer is full of pocket gophers, rats, and mice, which she disembowels before serving to Alice. During mating season, Alice will hoot the night through, which makes for spotty sleep. "And then you've got those razor-sharp talons," she says. "If she has some cached food in her room that I am not aware of and I get too close, she runs over and pounces on my feet. Normally she doesn't grab hard, but it's still enough to poke holes. Also, she's very territorial, and if someone she doesn't know enters her room, she goes into attack mode."

This is no minor issue. Naturalist Ernest Thompson Seton called these owls "winged tigers." The scientific name for the Great Horned Owl is *Bubo* from the Latin for "horned or hooting owl," and *virginianus* because it was first spotted by western naturalists in the Virginia colonies. But the bird and its close relatives range widely, from the southern tip of South America north to the Arctic. And everywhere it lives, from farmlands and high plains to deserts, rainforests, mountains, and wetlands, it's a ferocious predator, capable of killing birds as big as an adult male peacock. Banders climbing trees to reach an owl nest have been struck hard by attacking adults.

Bloem is careful whom she brings into contact with Alice and always introduces them from a distance.

Once Bloem and Alice had settled into a cordial, or at least a tolerant, relationship, their close contact allowed Bloem to collect intimate observations on Alice's vocalizations.

One obvious question: Would a human-imprinted owl have normal vocalizations?

"I was pretty sure she would," says Bloem. "Owl vocalizations are generally accepted as genetically inherited, not learned," so Alice's hoots and calls were likely a programmed part of her behavioral repertoire. Unlike songbirds, which learn their songs through vocal learning, just as we learn to speak—by listening, imitating, and practicing—owl calls and songs are hardwired, genetically fixed. Researchers have taken owl eggs and put them in an incubator and then played the songs of song sparrows and other kinds of birds to them, David Johnson told me. As soon as the owlets are adult enough to call, they always sing "owl," the calls of their species.

But Bloem wasn't sure Alice would deliver her calls in the appropriate behavioral context. So she decided to supplement and confirm her observations of Alice by listening to the calls of wild owls in the forests around her and watching their behaviors with infrared night vision binoculars. However, she found that she couldn't see what was going on around the nest. "Here, Great Horned Owls tend to nest where there are trees," she says, "and where there are trees, there are branches, and infrared illumination produces glare off branches." To observe their behaviors while vocalizing in a breeding context, she realized she needed a captive breeding pair.

Enter Rusty and Iris, injured adult Great Horned Owls who joined the International Owl Center in 2010. Both birds had eye issues. Iris had a punctured pupil, and Rusty was blinded in a car accident. Neither were human imprinted. "They met in rehab and were hanging out together," says Bloem, "so they were a good potential breeding pair for us." The center built a breeding facility outfitted with security cameras and microphones and then livestreamed the pair on a webcam on the internet so people watching could help with observations of the pair and their offspring.

Finally, Bloem had everything in place to conduct her study.

The calls of Great Horned Owls are "varied and difficult to characterize," according to the authoritative *Birds of the World*. A male Great Horned Owl can produce "an indescribable assemblage of hoots, chuckles, screeches, and squawks," wrote one ornithologist, "given so rapidly and disconnectedly that the effect is both startling and amusing." The owls are known for their deep, resonant hoots in a quiet baritone, which often contribute to a midnight mood in movie soundtracks.

After hundreds of hours of meticulous observation, Bloem managed to characterize and describe fifteen separate vocalizations: six sorts of hoots, four types of chitters, and five kinds of squawks, including an alarm squawk like an eerie shriek. She also noted that the owls have nonvocal communication. When they're fearful or agitated, they'll hiss or clack their bills.

Like most owls, Great Horned Owls are highly territorial. They perch at the edge of their territory and establish its bounds with their voices, uttering deep, soft hoots in a stuttering rhythm: *hoo-h'HOO-hoo-hoo*. Both males and females give territorial hoots on their own, singing simultaneously with a mate, or even dueling with a neighbor or stranger owl. It's far better than an actual physical battle. "Owls don't want to fight because the risk of injury is so high," says David Johnson. "If you catch a talon in the eye or something, it's game over, so you're willing to do everything you can do to avoid that. When you vocalize, if you want to be a tough guy, you'll lower your voice and project it."

A low-frequency hoot carries well and allows for maximum reach with minimum attenuation in a variety of habitats, assuming the hooting isn't too close to the ground. When my friend Kinari Webb was studying orangutans in the rainforests of Borneo, she and her colleagues used owllike hoots to signal one another. "A medium-pitched *whoo-whoo* call carries well," she says, so it's the method most Indonesians use to locate one another in the forest. From the right perch, even the subtle hoot of the Long-eared Owl can travel more than a third of a mile, if wind or traffic doesn't drown it out.

In a full-blown territorial hoot, a Great Horned Owl leans forward in a nearly horizontal position, plumage bristling, throat puffed out like a giant frog, tail cocked up. While female Great Horned Owls are considerably larger than males, they have a smaller syrinx, which gives them a higher-pitched hoot. Bloem begs to differ with the conventional view that males have a more elaborate hoot than females. In fact, she says, a typical male hoot has only four or five notes, while a female hoot has more like six to nine. In fairness, males do occasionally let loose with an impressive vibrato on their second note, which might account for the male complexity bias, she says, but that's uncommon.

The chitters range from clucky, delivered in or near the nest or while standing on food, to annoyed and even screaming. The birds will give the annoyed chitter, a loud, high-pitched squeal sometimes accompanied by a sharp bite, when they're—well—annoyed. Then there's the full-on screaming chitter, when they're extremely irritated or being physically restrained.

Hungry young owls beg for food with a harsh, screeching squawk. Females away from their young make this call, too. It's a sound that's easy to locate and effective in communicating over short distances. When Alice is bored, she will "double squawk," says Bloem, a sort of *wac-wac*, to get attention. Bloem also noted a specific vocalization Iris made while feeding her chicks to stimulate them to eat. Later, when Bloem hand reared owlets herself, she found that if she offered them food at a time when they just were not interested, all she had to do was play that call "and all of a sudden they wanted to eat."

Another kind of squeal she had never heard herself was brought to her attention at a conference by two researchers who work in Saskatchewan. When the researchers heard her presentation on owl vocalizations, they asked if she had ever heard the calls that nesting Great Horned Owls deliver during a broken-wing display.

Great Horned Owls have a broken-wing display? This remarkable nest defense is typical of vulnerable ground-nesting birds like plovers and sandpipers. That Great Horned Owls, tree nesters near the top of the food chain, might resort to this sort of distractive display was news to Bloem. So she went to Saskatchewan to see for herself. And sure enough, if a nesting Great Horned Owl is threatened by a dog or other predator, it will fluff up its feathers and throw itself to the ground, flapping around as if its wing is injured and squealing once or twice—a highly risky move and an indicator of what good nest protectors these birds can be.

Some vocalizations, the more intimate ones that might occur between male and female members of a pair, Bloem learned only from close contact with Alice. The soft low-pitched "hello" hoot, for instance. Or the quiet, conversational chitter directed to another individual at close range and delivered in short bursts roughly transcribed as *Hmm? Hmm? Hmm? Hmm?*

A whole bucket of vocalizations occurs in the context of copulation. These Rusty obligingly demonstrated again and again before he sadly passed away in 2022. "Anybody who watched our webcam knew when Rusty was interested, because his hoots got lower and slower," says Bloem. "And sometimes he left notes off at the end. He would sit across the aviary and stare at Iris or sometimes sidle over and hoot right in her ear, softly, *Hoo, Hoo,* and then just keep going until she responded with a little high-pitched hoot. Then he would get excited and utter these staccato hoots, almost like a chimpanzee, and fly in, land on her back. She would move her tail aside, and he'd wrap his tail around her for what's called the 'cloacal kiss,' when semen is transferred. If he was successful—or even close to successful—he'd let out a high-pitched screech, the 'copulatory squeal.'

"You would think that sound would come from the female because she's got eight talons in her back," Bloem says. "But she's busy doing those annoyance chitters. I had to watch a zillion copulation sequences before I could sort this out. Finally, Iris actually hooted during one of the copulation squeals, so then I knew it was Rusty who was doing the squealing."

Over time, Bloem established the complete vocal repertoire of Great Horned Owls and linked the various calls with the birds' associated behaviors. But she wasn't done. When Rusty and Iris started breeding, she realized she could also study how owls develop their hoots and other calls.

To her great delight, she discovered that owlets begin vocalizing in the egg, even before they hatch. "We have a supersensitive microphone less than four feet from where the eggs were laid," she says, "and we were able to pick up the sounds of the owlets in the egg about two days before they hatched. They break into the air cell in their egg and start breathing air. That's when they start vocalizing. You can actually hear the little chitters of the chicks in the egg."

At just a little over two weeks, the tiny owlets hooted in the correct body position and with the correct rhythm—also a surprise. "Somebody who was watching the ball of three fluffy babies on the webcam sent me a note saying, 'Hey, I think the babies just hooted,'" recalls Bloem. "I'm like, '*What?*' And I took a look at the video recording myself, and there they were, those tiny little kids, in the proper position—heads lowered, featherless tail stubs tilted upward—hooting in the correct four-note rhythm, albeit in very squeaky little voices. Holy cow, nobody had any idea they could hoot when they're only two weeks old!"

The owlets continued their little hootlets for a few weeks and then stopped. At five months, they started up again, with voices that cracked like the voice of a teenage boy. Around six or seven months, their hoots became indistinguishable from those of adults.

The owlets also hooted in the correct context from the start. "This one took me totally off guard," Bloem recalls. "I was watching a brood, and I could see that the owls were obviously upset and alarmed about something. But they were all quiet and they were all looking outside. And suddenly, I hear this little high-pitched sound, and I'm thinking, 'What the heck is that?' Then the camera moves to show the back of an owlet named Patrick, and I notice that his shoulders are moving in sync with the sound. Patrick was doing an alarm call in his high, squeaky voice. I had no idea that juveniles would make alarm calls.

"And there was no practicing. The owlets simply produced hoots of the proper rhythm in the proper position and context from the beginning. They just got it right from the get-go."

If the vocalizations of owls are hardwired, does that mean one owl's territorial call sounds similar to another's? Or are owl voices like ours, rich with signatures of individual identity?

Some years ago, a team of French scientists set up an experiment designed to see whether Little Owls could discriminate between the territorial hoots of neighbors and strangers.

Like other owls, male Little Owls hoot to defend their territory, and they tend to hoot from a usual perch. If they hear an intruder hooting in their territory, they'll respond vociferously, with stepped-up hooting and combative flights. The research team played the hoots of a familiar neighbor and a stranger from two locations—from the usual perch of the neighbor and from an unusual location, some distance from that perch. No question, the Little Owls could tell who was who. When the hoots of a familiar neighbor were played from the usual location, the owls showed little response. This reduced aggressiveness toward neighbors is known as the "dear enemy" phenomenon. It saves an owl time and the energetic costs of signaling, patrolling, and chasing. However, when the researchers swapped in a stranger's hoots played from that familiar location, the owls quickly responded with a burst of shrill, excited, catlike hooting.

"We see this with our captive owls when a new wild owl shows up in the area," says Bloem. "First the captive owls are quiet and listen for a bit, then they really go into high gear hooting and are clearly very worked up. I've used the hooting of our captive owls to help me decide if the owl I'm hearing is a new one or one that has been in the area." She found that she could identify individual wild owls living in the woods around her by their territorial hoots, which were consistent for each bird and sufficiently distinct from one another to "fingerprint" individuals. What typically differs from one Great Horned Owl to the next is the number of notes per hoot and how they're spaced. The owl she named Scarlett Owl'Hara (owl researchers seem to love puns) sang a doublet of notes in part of her hoot. Wheezy sang a triplet, and Ruby's hoot rose and fell in pitch. Victor had a sexy vibrato.

Knowing the owls' individual voices gave Bloem an intimate window into their "love lives." Victor paired with Virginia, Wendell with Wheezy, Jack with Jill, Haggar with Helga, and Scarlett—eventually—with her Rhett, who later paired with an owl named Delilah. To Bloem's surprise, their pair-bonds were constantly evolving, with so much mate switching it was hard to keep up. "This isn't supposed to happen!" she exclaims.

Even Alice isn't immune to the vagaries of affection. When Bloem remarried, Alice switched her allegiance to Bloem's new husband, Hein. "Now Hein is number one, and I am clearly number two," she says. "So I kind of feel like she's a bit of a traitor. But you don't argue with a female Great Horned Owl. She is boss."

Each individual Great Horned Owl has a signature hoot. To varying degrees, this is true for all owl species, from Barred Owls and Tawny Owls to pygmy owls and screech owls.

Why would owls have distinctive voices? Most bird species have at least some individuality in their calls and songs. It's useful in identifying kin and communicating with mates, allies, and rivals. "Owls have it for very good reason," says Bloem. "After all, they interact mainly through vocal communication over multiple seasons and over years." It makes sense that their voices would be distinct and stable, to recognize one another and maintain long-term pair-bonds, to reunite in consecutive breeding seasons, to know who their neighbors are. "As long as a hooting owl is familiar and staying in its own territory, there's no need to get bent out of shape," she says. "But you want to know if a total stranger shows up so you can crank up the hooting to let them know your spot is taken and hopefully make them leave before any physical altercation erupts. Owls decide how to respond to other owls hooting near their territory based on this recognition of individuals."

How owls themselves recognize one another by voice is still largely a mystery. But lately progress has come from careful observation and from machine learning and computational science aimed at understanding how an owl's hoot might convey a caller's unique identity.

Pavel Linhart, a behavioral ecologist at the University of South Bohemia in the Czech Republic, studies vocal communication in animals ranging from piglets and pipits to willow warblers and owls to try to understand how their hoots and squeals and warbling songs communicate information about body size, emotional state, and individual identity. To study vocal individuality, researchers select and measure acoustic features of calls, such as duration, frequency, and what they call spectral features—sound quality and "color"—and how these change over the course of the call. Then they use algorithms to build a system to classify the calls. In this way, they can calculate how many unique individual signatures are possible for a given population or species and how easy or difficult it is to assign a call to an individual based on its acoustic features. Overall, owls as a group have high individuality compared with other animals, says Linhart. This makes sense given the role of sound communication in their lives.

Linhart and his colleagues have created an interactive web presentation of vocal individuality in owl species so that people can hear for themselves what makes individuals within a species stand apart (high or low hooting, melody, rhythm, or sound quality—tonal or raspy). He points out major differences in the calls of different species. Barn owls typically use a broadband screech for their territorial advertisement calls, which vary from bird to bird in how the screech rises in pitch and how suddenly it ends. In other kinds of owls, individuals' calls vary in the melody or frequency (higher- or lower-pitched hoot). In screech owls and the *Bubo* genus of owls (eagle owls and Great Horned Owls), the rhythm of repeated units is the distinguishing feature. Species in the *Strix* genus (Barred Owls, Tawny Owls, wood owls) seem to have the most complex hooting patterns, including several different syllable types, and highly individualistic modulation and rhythm. The prize for highest individuality of territorial calls goes to Tawny Owls.

Why some species have more individuality in their calls than others is a puzzle, but Linhart thinks that species with more complex calls encode more individuality, and that, in fact, complex calls may have evolved to allow better individual discrimination.

Lately, Linhart has explored individual variation in the calls of Little Owls. These owls live in the open country, fields, pastureland, and rocky hillsides of western Europe all the way east to Russia and south to Mali and Niger. In the Czech Republic, they were once a common species. Today the population there is considered endangered. "Now more than ever for the Czech population of Little Owls, each individual matters," he says. "And now more than ever, understanding the unique sounds of individual birds matters for monitoring and conservation."

Because Little Owls are sedentary—they don't migrate—and have stable territories that they occupy long term, they're a perfect model species for studying acoustic signature identification, says Linhart. In the past decade, he and his colleague Martin Šálek have cataloged the owls' territorial voices. Now they, along with PhD student Malavika Madhavan, are investigating whether Little Owls that breed in high densities have more distinctive territorial calls because they "need" to be more recognizable to others. That is, do individuals with many neighbors have more highly individualistic voices?

The team members studied various regions with different densities of Little Owl populations. In Hungary, they found that where males are living on isolated farms, far from other neighbors, territorial calls have less individuality. But where the population is denser, where there are as many as five males at a single site calling to one another as little as one hundred meters apart, calls are more distinct and distinguishable—likely to enhance individual recognition within the community.

To illustrate the variation in Little Owl hoots and how hard it is to hear them, Linhart has created a series of recordings and posted them online: First, the calls of three males are played at normal speed

Little Owls

and then the same calls are played at a speed three times slower than normal so listeners can hear the nuances, the differences in the way individual birds modulate the frequency of their calls. Linhart has also devised an online game that demonstrates—with humiliating efficacy—the difficulties of recognizing the different calls of individual Little Owls. It's a game of "match the pairs" with sixteen "cards" made of audio buttons. Each card is the hoot of a different Little Owl. The challenge is finding matching hoots among the cards. I've played the Little Owl game again and again, and each time I fail miserably. I simply cannot hear, or cannot remember, the distinctions in frequency and pitch of the vocalizations. It makes me admire even more the talent for detecting subtleties in owl hooting that owls—and some owl researchers—seem to possess.

Linhart told me that most people fail at the game. Little Owls have a lower level of individuality in their calls than do other species. Many males have very similar calls. "It's very likely that owls could hear faint differences in pitch and frequency modulation or duration, but these details are not perceptible to us." Also, Little Owl hooting can sound different at the beginning of a calling sequence than it does at the end—lower and slower at the start, faster and higher toward the end—like the male is just warming up or finding confidence. Finally, some Little Owl males have a variable calling style, which could be a function of age or social environment. (Older males might be more consistent and less intimidated by other calling males.)

I'm beginning to feel a little less humiliated.

The point is, telling apart the signature hoots of individual owls isn't easy. It takes sharp hearing, familiarity, and hours and hours of listening. It also helps to have a musical ear.

When Karla Bloem was studying the calls of wild Great Horned Owls, she got lucky when Marjon Savelsberg showed up as a virtual volunteer to help with the monitoring. A musician by training, Savelsberg specialized in baroque music before she fell in love with owl calls. "She was magnificent at telling apart the wild owls and hearing extremely distant things," says Bloem. "She was perfect for it, a genius. Her brother is a rocket scientist. She is the equivalent in the world of owl vocalizations."

In recent years, Savelsberg has tuned her ear to the monitoring of Eurasian Eagle Owl populations in her native Netherlands. Her discoveries have illuminated some of the mysteries of these highly secretive owls and the intriguing (and somewhat scandalous) details of their family lives.

Twilight in the vast limestone quarry of Oehoevallei (Eagle Owl Valley) near Maastricht, an old Roman city in the southern Netherlands, now a busy metropolis. The land around the quarry is green and rolling, with oak, beech, and fruit trees and innumerable hidden wonders, rare plants and animals that thrive in the chalky grasslands—orchids and other wildflowers, and the unusual insects drawn to them, including endangered butterflies and twenty-four species of dragonflies. The quarry dates to the thirteenth century, when people discovered that the marl, the yellow chalk stone of the region, was good for building houses. Beneath the rolling grasslands and woodlands is a network of human-made caves and tunnels, where workers once sawed out blocks of chalk. Because of its special vegetation, the quarry has been set aside as a nature area by the conservation organization Natuurmonumenten. During the COVID-19 pandemic, the place became a refuge for city dwellers seeking green space and safe recreation.

But now the day is growing late, and picnickers and hikers are winding up their outings and returning to their cars. The quarry settles into quiet. It begins to drizzle.

Then, suddenly, the low rumble of a small engine, and Marjon Savelsberg rolls along a path on a mobility scooter up to a locked gate, stops, and lets herself in with a key. She motors down to her chosen spot and then backs her scooter deep into a bush. The air is thick with herbal smells, she says, "like a kitchen cabinet full of spices." She pulls out a microphone and a small handheld recorder and settles in for a night of observation and recording.

It's about an hour before sunset, "just when the eagle owls get busy," she says. "They wake up and come out of their roosting spot and you see them stretch their wings and stretch their legs and shake their feathers. They perch while it slowly gets dark, and then they start to call or fly. I just sit there in the shrubbery where they can't see me, and watch and record."

All is quiet at first, just the patter of rain and the high-pitched *beep beeps* of midwife toads and an occasional call from a Common Cuckoo, just like the clock for which it's named. But then, overhead in the evening air is a deep, booming *Ooo, hoo*. And from afar, another lower, longer call, *Ooo hoo hoooooo*. Two males, one with a call a fraction higher than the other. Savelsberg has no trouble telling the two individuals apart.

The name of the Eurasian Eagle Owl evokes a hybrid mythical creature, half eagle, half owl, and that's not too far from the truth. At ten pounds or more, with a wingspan of more than six feet, the bird vies for size with the Blakiston's Fish Owl, with eyes like a fiery sunset that stare down, imperious, on mere mortals below. The most powerful hunter of all owls, it's capable of taking pretty much anything it pleases—rabbits, geese, coots, foxes, even roe deer—surprising prey by flying close to the ground or treetops or seizing birds and bats in full flight. It's also notorious for feeding on its predatory cousins—hawks, buzzards, and other diurnal raptors—sometimes systematically searching rock crevices and snatching a Northern Goshawk or Eurasian Hobby from its nighttime roost. "We've found Peregrine Falcon bands in their nests," says Savelsberg, "along with the remains of coots, geese, Long-eared Owls. You name it, they eat it." They will even eat a hedgehog after peeling its spiny skin.

For power and ferocity, this bird has no rivals and yet—as Savelsberg will tell you—it also displays an extraordinary kind of tenderness. When she first handled a Eurasian Eagle Owl chick to measure, weigh, and band it, she was struck by its legs. "I never realized what an owl leg would feel like," she says. "You don't think about those things. But even when they're very young, four weeks old, and they can't really walk or fly, their legs are so big and sturdy, all the way up to the thighs. It's pure muscle, muscle covered in velvet, so soft and strong at the same time. That's what this bird is for me, incredibly soft and strong. You see it with the big mothers, weighing two, three, four kilograms, caring for these tiny little chicks weighing fifty grams. The tenderness they have is wonderful and yet with those feet they kill rabbits without hesitation."

Eurasian Eagle Owls inhabit a vast area, about twelve million square miles, from western Europe to the Far East, largely because they can adapt to a wide variety of climatic conditions, habitats, and altitudes. However, human pressures have taken their toll—shooting, poisoning, habitat destruction, collisions with cars and trains, and high-voltage wires—and in the twentieth century, the numbers of these magnificent birds dwindled.

"In my country, the Eurasian Eagle Owl was gone for a century, eradicated by persecution and hunting," says Savelsberg. In Europe, the numbers had fallen so low in several countries that Germany started a massive reintroduction program in the 1970s. It was so successful that the birds started moving back into the Netherlands. Now the province around Maastricht has twenty-three pairs.

In 1997, the first pair of breeding Eurasian Eagle Owls was found in the Maastricht quarry. The birds like to nest on sheltered cliff ledges or in crevices or cave entrances, so Natuurmonumenten created nooks in the quarry walls for nesting sites. "Nothing can reach the nest there, so it's a fantastic place," says Savelsberg. "It does get hot in summer, but apparently the females don't mind. The males always sit on the opposing quarry wall and keep an eye on things." Now the quarry site is home to three different pairs—a fact known only because of Savelsberg's keen ear and ability to tell one bird apart from another by slight differences in their calls.

By all accounts, monitoring these birds is extremely challenging. The species is nocturnal, difficult to spot because of its cryptic plumage

Female Eurasian Eagle Owl at Oehoevallei

and excellent camouflage, with little visible distinction between the sexes. Knowing the calls of individual birds is key to the task.

In 2008, a team of French scientists analyzed the vocalizations of a small population of Eurasian Eagle Owls breeding in the Loire Valley and found that the calls of both sexes were individually distinctive and that monitoring the owls by analyzing their individual vocal signatures could help in understanding their populations.

"Pitch, or frequency, tells a lot about sex," says Savelsberg, "but in order to know what you are hearing when an owl calls, your ear must be able to perceive pitch, and the brain must be able to remember it, and for me that is a very natural thing to do.

"Eurasian Eagle Owl vocalizations are music to my ears," she says. "Of course, they don't have the melodic variations of many songbirds, but they do have quite the repertoire. Do you know 'Continuum' by György Ligeti?" she asks me. "It's a piece for harpsichord, and the musical tension is amazing because of the small intervals and variations. Eurasian Eagle Owl vocalizations are like that. The differences in hoots can be minimal, but they're there, and they fill me with wonder. Just the way music does."

Like Bloem, Savelsberg did not intend to study owls. A classical musician by training, she studied with members of the Johann Strauss Orchestra and hoped to become a professional recorder player. But just before her final exam to finish conservatory, she developed difficulty breathing and controlling her muscles and couldn't play her instrument for the required hour. This forced her to take a different path.

"I had to give up being a musician," she says. "Well, give up being a musician *physically*. Up *here*"—she taps her head—"I am a musician forever." She retrained as a teacher and, for a time, taught young children with disabilities, but then had to give that up, too, when her condition deteriorated. The doctors told her she had idiopathic cardiomyopathy—a disease of the heart muscle that makes it harder for the heart to pump blood to the rest of the body—and that she would probably not live more than another decade. "It was hard. I went down for years," she recalls. "But then I said, 'Okay, I'm here now. Let's see what I can do.'

"I've always had a fascination for nature," she says. "Even when I was a small child, I had a little stroller, supposedly for my dolls. But once a week, my mom would turn it upside down and out would fall feathers and pebbles and other stuff I'd collected." She found a webcam run by the Dutch Little Owl Working Group, an organization in the Netherlands that was looking for webcam viewers to record the prey delivered to a nesting pair of Little Owls. She jumped in to help with the observations. Later, in 2012, she discovered the webcam for the Great Horned Owl vocalization project at the International Owl Center and became one of the chief observers on that project.

"She was so inquisitive," says Bloem. "She started studying the literature and analyzing the spectrograms of the vocalizations and really became a partner in it all."

"I became very interested in reading spectrograms because they're really the music scores of nature," says Savelsberg. "They're almost like sheet music, where the high notes are the soprano part and lower down are the bass notes, and the shape of the note tells you how long a sound lasts. In spectrograms, it's basically the same. Both are graphical representations of sound. As a musician, I used to like sitting on the train reading sheet music and hearing the violins and other instruments in my head. Now I can do that with spectrograms, with the vocalizations of birds, robins, whippoorwills, owls. I don't even have to hear their song and calls anymore. I know what they look like on the spectrogram and can hear them in my head."

Savelsberg's curiosity drove her to visit the local quarry near Maastricht to see whether she could hear the Eurasian Eagle Owls there. She knew there was a pair at the site, though she'd never seen the birds, and wondered if they sounded similar to Great Horned Owls. It was the beginning of a new career.

"I went from a wooden flute to a handheld recorder and a shotgun microphone," she says. "I couldn't walk very far, so I got a mobility scooter and went to the quarry every night to make audio recordings."

One night, a forester from Natuurmonumenten saw her and asked what she was doing there in the dark. "I told him, 'I'm a bit of an owl fan, and I'm recording their vocalizations, but they don't really hoot during the day.' And then he said, 'Well, you should contact our ecologist.'" She did. Now she's considered an expert on Eurasian Eagle Owls and their vocalizations and works closely with the park ecologists to band the owls and monitor the population.

"Unbelievable," she says. "From that moment on I was allowed to go into areas where no one else could go. I suddenly had colleagues again and was seen as much more than someone with a disability. After having to give up playing music, I had been so depressed. It was my passion in life. I wasn't able to listen to music for ten years. I couldn't—it hurt. And then I discovered this work and realized I was still a musician. All the skills that I learned, all the talent I have, I can still use, just in a different way. Because I'm so fascinated by sound, I can do a lot for this species."

Sitting still in the dark quarry, tucked into the bush and nearly invisible, Savelsberg listens.

Both male and female Eurasian Eagle Owls have a deep, sonorous, booming two-syllable territorial hoot, a loud *ooo* and lower, descending *hoo*, but the notes are often so close they sound like a single slurred hoot, repeated every ten seconds or so. "They sing from a prominent perch," says Savelsberg, and as they sing, they fluff out and flash their little white rectangular throat patches, which are thought to be signals of a male's physical condition. The male with the brightest badge often has the best territory, and the owls may even hoot in a strict hierarchical order based on this status. Here's one reason for an owl's ultraviolet light sensitivity. The white feathers on that throat badge reflect ultraviolet light and appear whiter and brighter to other eagle owls, thanks to their ability to see UV light. Fledglings have the same sort of patch, but in their mouths—visible only when they're gaping or begging for food. Both patches flash bright—on the throat, to help the adult eagle owl beckon its mate or warn a rival that might intrude on its territory, and in the mouth, to help the chick keen for a parental snack. The throat badges may be the reason these owls hoot primarily at dusk: twilight heightens the visual contrast of the white patch.

The different varieties of eagle owl vocalizations resemble those of Great Horned Owls and other species—hoots, squawks, chitters, soliciting and begging calls. But if you wander too close to an eagle owl nest, says Savelsberg, you may hear a call unique to this owl that will lift the hair on your neck, a chilling, raucous laugh-like "devil's cackle." The owl sometimes uses it as an alarm call. "But I think it has more meanings than this," she says. She has watched females devil cackling for forty-five minutes at a time in the autumn when their owlets are dispersing. Young owlets only three months old will cackle, too. She has also recorded a few vocalizations that have not yet been described in the literature, including a contact call used by both males and females to stay in touch.

In the old days, Savelsberg would have to lug around a battery pack weighing sixty-five pounds. Now the equipment is lighter, easier to transport on her mobility scooter. She also places tiny recorders in busy areas of the quarry, fixing them under a leaf so no one can find them. Then she drives around, scooter bouncing up and down the hills, checking out the owl activities in the dark.

She doesn't worry about being alone in the quarry at night. "People always say, 'Oh, I wouldn't dare to go there after dusk.' But really, it's a second home to me." She did once encounter someone in the dark, a man climbing the quarry wall right where she was sitting. "But he was as freaked out by me as I was by him—maybe more so."

The only time she ever lost equipment to theft, the thieves were not human.

She had been observing a pair of owls in a remote area of the quarry. "This pair has a fairly small territory, so I could get very close to the nest," she says. "And I found a place where I can observe them without disturbing them. I've spent hours there watching what's going on. I can even hear the owlets when they are just out of the egg. It's fantastic."

One day she decided to leave her recording equipment in place at the spot. The next day, she came back to retrieve it. "Everything was gone except for the tripod. Microphone gone, DeadCat gone, cable gone, power bank gone, recorder gone. I thought, 'Oh, no, it was all stolen.' I went away so unhappy. But in the days after, something kept nagging at me. Why would someone leave the tripod there?

"After a few weeks I went back, and under the owls' favorite roosting tree, I saw a package hanging high in the bushes that looked like my recorder and my power bank, still all wrapped in plastic. The microphone had dropped one hundred thirty feet from the tree to the quarry wall, but the DeadCat was gone." To be fair, she says, the fuzzy DeadCat does look a lot like a furry rodent. "Yeah, but those stinkers stole my equipment, so I had to call them Bonnie and Clyde. Now I often put my recording setup under an umbrella, but it's not much protection. The owlets jump on the umbrella and tear it apart. Always a surprise what I find!"

After Savelsberg pulls the little memory cards from her recorders, she loads them onto a computer at home and starts to analyze what she has recorded. She has trained sound-analysis software called Kaleidoscope to recognize the hoots of Eurasian Eagle Owls on the recordings. "At the beginning, it would 'recognize' lawn mowers, kids playing football, the call of a Common Cuckoo," she says. In fact, the software is still confused by the cuckoo's hoot-like *cuck-oo, cuck-oo.* But she works through the recordings methodically—it's a huge amount of data, eight terabytes per year—to separate out the owl vocalizations and assign them to individuals, then correlate the vocalizations with the behaviors she sees.

"The fantastic thing about individual vocal recognition and acoustic monitoring is you can count how many owls there are," she says. "If an owl disappears, and a new one comes in, you can hear it, and you can see it in the spectrogram. So you have scientific proof to validate what you think that you're hearing.

"It's such a rewarding way to monitor owls. Knowing the unique sounds of individual owl voices means that you can accurately monitor them without having to capture or trap them and put transmitters on them, so you don't have to disturb them or cause them stress. You just go there, position your microphones, and listen. You don't even have to be present when you're recording. You're just sound observing the owls in their natural environment with their natural behavior."

Time, patience, a musical ear: it's a recipe for deep insight into how the owls use the valley; the boundaries of their territories and the posts from which males hoot at one another through the night; where they roost, hunt, build nests. And, of course, their population and their family life.

Until Savelsberg arrived on the scene, just a single pair of Eurasian Eagle Owls was thought to be living in the quarry. Now she knows by call the individual members of all three pairs and their territories and has given them names for use in her talks (in her data, they're identified by number). "Since it's a quarry, Flintstones is the theme," she says. "We have Betty and Barney. Fred and Wilma. Pebbles and Bamm-Bamm."

What she discovered while tracking the owls, their territories, and how they're paired has overturned some old notions about Eurasian Eagle Owl behavior. It was thought that pairs stayed together for life. Not so. "Tracking their shifting relationships is the coolest stuff you can do with individual recognition," says Savelsberg.

In the bird world, divorce is defined as one breeding individual partnering with a new mate while the previous mate is still alive. A study in barn owls suggests it occurs when breeding isn't going well for a pair, and new pairings benefit both partners. But not much is known about divorce in eagle owls. Savelsberg discovered that those at the quarry seemed to swap partners as often as Great Horned Owls.

"Sadly, Bamm-Bamm died one August, so Pebbles was alone, and she was hooting. By December she had found herself a new mate. He moved in, and—I almost couldn't believe my ears—it was Barney!

Barney had abandoned Betty and gone to live with Pebbles, and they formed a new pair and had owlets. Then, at the end of the breeding season, when the owls dispersed, Pebbles disappeared—after living with Barney for only a year. So poor Barney hooted and hooted and hooted, and not long after, guess who joined him? Betty. Betty left Fred to live with Barney. And then not much later, Fred found himself a Wilma.

"If you want soap opera," she says, "just learn individual recognition of eagle owls."

Savelsberg's data on Eurasian Eagle Owls and her insight into individual owls are being used for a project developing special software and machine-learning algorithms, like the ones Linhart uses, to detect the individually distinctive patterns of vocalizations in animals of different species. The project, led by the Naturalis Biodiversity Center in the Netherlands, harnesses machine listening and computation to analyze the voices of individual animals and the sounds they make in different conditions. Savelsberg provides the center with data sets of individuals vocalizing in a variety of circumstances (wind, rain, leaves rustling on trees) to test these programs. The goal is to understand the mysteries of animal vocalizations and to use the acoustic signatures of individual animals, including owls, to monitor populations, just as Savelsberg does—a potentially powerful tool in conservation.

In her nights at the quarry, Savelsberg has observed other revealing aspects of Eagle Owl social interactions. Parent owls don't chase away their young at the end of the breeding season, as was previously believed. The owlets hang around, sometimes for months. This developmental pattern—a long juvenile period before becoming independent from parents—is considered by some scientists a prerequisite for intelligence in birds.

Eurasian Eagle Owl chicks

Even more surprising, Savelsberg spotted a pair of adult eagle owls adopting an owlet, a six-month-old juvenile that was completely unrelated to them. The pair had lost their own chick at ten weeks. "The owlet was from another pair that had stopped feeding it, and it found its way to this neighboring, chickless pair," says Savelsberg. "Its siblings had gone to the other side of the quarry, but this owlet was moving closer and closer to the other adult pair. Suddenly, it sat itself down in the middle of their territory. And I thought, 'Are you nuts?' And it began its begging call, *raah, raah*. And then the pair starts feeding it! I couldn't believe my eyes. I went there every night for three weeks and just sat there and watched the female come to feed it."

It seems counterintuitive. But many bird species rear offspring that are not theirs, especially if they're captive bred or cooperative breeders (birds that routinely raise young collectively). Still, the behavior is unusual in wild owls. Young barn owls are known to switch nests if they aren't getting enough food from their own parents and are often folded into their adopted family. Owl researcher Sumio Yamamoto recently reported a case of a wild male Blakiston's Fish Owl helping to raise a chick that was not his. The biological father of the chick was injured, and the unrelated male "either knew of the original resident male's injury or, in fact, caused it," says Yamamoto. In any case, the male fed the nestling and helped to rear it, possibly to solidify his pair-bond with the resident female.

Sometimes owls will adopt orphan chicks when they lose their own, which appears to have been the case with these eagle owls. But the behavior seems remarkable, nonetheless. "I can imagine if you have four chicks of your own and then a fifth one appears, and you just fold it in," says Savelsberg. "But if you don't have owlets of your own, haven't for two and a half months already because yours didn't make it, and there's this big juvenile begging, and you still feed it?"

Owls defy expectations. They show us that their family lives—their pairing and unpairing, their parenting and raising of their young—are far richer and more complicated than we thought. Sometimes the only way to see this is to listen.

Five

What It Takes to Make an Owl

COURTING AND BREEDING

Northern Pygmy Owl chicks, a day or two out of the nest, photographed by Daniel J. Cox / naturalexposures.com

On a brilliant June morning at a farm in Linville, Virginia, the sky is blue, the grass is green, and birds are nesting everywhere: bluebirds and Tree Swallows in boxes, robins in a cherry tree loaded with ripe fruit, and low in the hanging branches of a gorgeous green weeping willow tree, Orchard Orioles in a dewdrop nest of perfection woven from dried grass, lined with soft plant down and feathers, and stuffed with three big chicks. The nest is dappled with light and swings gently in the breeze. From the limbs above come

Barn owl chicks in a silo at the Linville farm

the high-pitched whistle and occasional *psssht* of the mother. A Brown-headed Cowbird chick already fledged from this little nest. Its brood parasite mother chose well in dumping her egg there.

Up a path, in the farm's Civil War–era barn, a family of barn owls is nesting under very different conditions, in dim light on the hard floor of the old silo, its walls bedecked with whitewash, its floor packed with grungy black owl pellets several inches deep and no nesting material to speak of.

It's hard not to wonder at the contrast in nurseries. But in fact this is an excellent owlery. Barn owls have been nesting in this silo for at least the past twenty-five years, probably longer. Now five owlets sit hunched on the floor, all scrunched together, circling their heads and hissing. The youngest is still a bundle of fluffy white down, like two big fuzz balls stuck together. Three are in the muddling middle, bits of fluff still clinging to them. The fifth is perched on a ledge above the others, fully feathered and full of attitude, looking for all the world like a gargoyle with a heart for a face. The oldest of the six chicks has already fledged, found its way up fifty or sixty feet to the top of the silo, and launched out into the world. It seems nothing short of miraculous.

What does it take to make an owlet? It's the mystery and the enterprise for which all owls live. The different species do the job differently, but for all, it's hard work with tough odds. And for all, it starts with finding a mate.

HOOTING, TOOTING, AND CLAPPING FOR LOVE

Pattee Canyon, western Montana: Sunrise is not far off, but it's late March, and it's still cold and dark. In a beautiful mixed forest of ponderosa, larch, and fir known as Larch Camp, Steve Hiro sits at the base of a tree, listening in the dark. Fifteen or twenty feet above him is a tiny cavity harboring an owl nest. Out of the gloom comes the soft trill and single toot of an owl—a male, Hiro figures. Then it's quiet. After a moment, a squeaky, higher-pitched double toot, the female in response. Quiet. A few minutes later, the same exchange.

These are not the "advertising" calls of a bird looking for a mate. The pair has already bonded. These mellow toots and squeaks are what Hiro calls "soft talk," a quiet, intimate conversation between a pair of already bonded Northern Pygmy Owls. But they are the upshot of a long and mysterious courtship that Hiro is working to fathom, trying to understand how these birds form pairs, the conversations that unite them, how they collaborate to choose a nest and ultimately to raise their young, all the secrets of two owls coming together to make the next generation of their kind. To understand their courtship, he listens for the birds in the dark and then tracks the development of their pairing through their vocalizations.

When Hiro began his work with Northern Pygmy Owls in 2009, there was almost nothing known about how the owls court or form a pair or even when they launch the enterprise. "We thought these owls paired up in April," he says. "But then I started going out in the woods earlier and earlier in the year, listening to them and following them, and I found they were courting all the way back in February."

Northern Pygmy Owls are small, brown, round-headed woodland owls, secretive, solitary, and typically quiet, except during the breeding season, when their vocalizations come alive. Their territorial call is not like the call of other owls; it's more musical. When Theodore Roosevelt and naturalist John Burroughs first heard it, they couldn't believe the "queer un-owllike cry" was actually an owl. Wrote Burroughs, "It was such a sound as a boy might make by blowing in the neck of an empty bottle."

The first Northern Pygmy Owl I saw flew in and settled on a high branch of a pine tree, a plump little bird no bigger than a junco, knotlike on the limb and visible only through a spotting scope, but clearly very vocal, puffing out its throat with each rolling note. As Mary Oliver wrote, "It's not size but surge that tells us when we're in touch with something real."

Later, I would see one almost eye to eye, tooting and trilling from a pine tree just a few yards away. I was alerted to its presence by a flurry of jeering alarm calls and frantic activity from a mob of nuthatches and chickadees. This is always a good way to spot an owl. Listen for alarm calls from robins, nuthatches, titmice. The little birds have good reason to be upset. A fierce hunter, the pygmy owl snacks on songbirds. And it frequently preys on birds and mammals larger than itself, twice its size, so bigger birds like Bohemian Waxwings, Hairy Woodpeckers, and Thick-billed Longspurs are not safe. It will even bring in chipmunks and red squirrels.

I can see why these owls are known as "glaring gnomes." They even glare from behind, or so it seems. On the back of the Northern Pygmy Owl's head is a pair of dark, white-ringed feathered eyespots, "false eyes" that are quite convincing. For years it was thought these eyespots functioned solely to confuse predators, but research suggests they may also confound mobbing songbirds.

Whatever may be said about the ferocity of these owls, no one can accuse them of being untender to their mates.

Early in the breeding season, in late winter, males sing their strange, loud, repetitive toots from a high perch to attract a mate. If the female shows interest, then comes duetting, alternating single toots from a distance, male at a lower pitch and female at a higher pitch.

"As the pairing progresses, the change in vocalizations is so cool," says Hiro. "Sometimes after the duet, you'll hear them copulate. There's a specific chitter so you know what's going on, a trill, but softer and faster, that only lasts for a couple of seconds."

When the pair is bonded, they'll hunt apart during the day, but they always know where the other is by swapping toots of call-and-response. And once they find a nesting site, then begins the soft talk—that quiet, intimate hooting around the nest site—as well as mutual preening, nuzzling, sharing food. "The male shows up at first light, and the first thing you hear is his mellow double toot," says Hiro. "If the female is in the nesting cavity, that might be her sign to come out and take a break from the eggs. They may or may not copulate. It's very intimate and it's haunting. You wouldn't hear it if you weren't sitting there, close to the nest, listening."

Listening is essential to monitoring breeding in these tiny owls. Their movements are so quick that you can blink and miss the female going in and out of a nesting cavity. Once she's in there, she keeps the location secret by rarely poking her head out. Hence Hiro's vigilant ear.

Hiro is a volunteer for the Owl Research Institute, an organization dedicated to long-term research on owls. It was founded by Denver

Holt on the Flathead Indian Reservation in 1987/88 and is now housed in a farmhouse near the small town of Charlo, Montana, in the shadow of the Mission Mountains. This part of Montana supports fifteen different owl species; fourteen of them breed here. Short-eared Owls nest in the grasslands and, in late winter, turn out by the dozens to search for voles in the evening. Long-eared Owls nest and roost in the draws dense with hawthorn, Russian Olive, and Chokecherry, and Boreal Owls breed in the high-elevation Engelmann Spruce and fir forests that cloak the mountains. Great Gray Owls, too, nest in the trees here and on early mornings populate the fenceposts on logged-over stump fields, hunting mice, voles, and rats. Even Snowy Owls make a showing in winter, perching on rooftops and fields around Charlo and south of Flathead Lake.

Holt's passion for owls began while he was an undergraduate studying wildlife biology at the University of Montana. A friend spotted the nest of a Northern Pygmy Owl not far from the university, and Holt and fellow student Bill Norton decided that learning about those owls beat out college. At the time, there was only a single reported observation of nesting Northern Pygmy Owls in North America, dating from 1926. The two undergrads skipped class and monitored the nest with binoculars and spotting scopes. They collected pellets and the remains of mice, voles, chickadees, flycatchers, nuthatches. They logged prey deliveries and recorded the owls' calls and songs. They noted that the male never enters the nesting cavity but, each evening at dusk, calls with a slow, hollow *toot-toot* to summon the female for feeding. They observed that the birds stick out their tails at a perky angle and twitch or jerk them from side to side, like a wagtail or a pipit, in excitement or as a kind of threat display.

The pair published two papers, and Holt never looked back. Now, almost four decades later, he knows owls as well as anyone on the planet. When I asked ornithologists for suggestions of owl experts,

Holt's name came up again and again. According to John Fitzpatrick of the Cornell Lab of Ornithology, he's "Mr. Owl." Around Charlo, he's known simply as "that owl guy" and has gathered around him a menagerie of eager supporters and staff, among them ranchers, farmers, bikers, professional hockey and football players, and young aspiring wildlife biologists. Seasonal employees, interns, and a slew of volunteers contribute hundreds of hours every year to the ORI projects. "His enthusiasm is infectious," one volunteer told me. Holt's baseball cap reads "umiaq"—the Inuit word for a large open skin boat that takes a crew to row.

Over the past four decades, Holt and his team have studied breeding Northern Pygmy, Northern Saw-whet, Boreal, Western Screech, Great Gray, Flammulated, Short-eared, Long-eared, Northern Hawk, Burrowing, and barn owls here in Montana. Every summer, he migrates to the Arctic to study breeding Snowy Owls. His goal in all this research is to fathom what owls need for their successful reproduction—food, habitat, nest sites—so wildlife and land managers can better work to conserve them.

I've joined the ORI team during the nesting season to observe some of their work and learn what goes into the making of an owlet. The institute's efforts to understand the breeding of owls begins with the kind of research Hiro does, finding courting birds. Hiro has been at this work since 1995 and came to it because of Holt. A retired heart surgeon, he specialized in complex cutting-edge mitral and aortic valve surgery and was turned on to owls when he bid on an auction item at a hospital fundraiser, "A Day in the Field" with Denver Holt. "I told my wife, Terry, 'I'm doing this, I don't care what it costs.'" Hiro won the bid and in November of that year went out in the field with the ORI team to band Long-eared Owls. "Hey," Holt said at one point during the day, "you're good with your hands. What do you do?"

Hiro discovered that he loved owls. He loved the process of field research and working with the ORI crew. After that, he volunteered whenever he could. The owl work was a welcome relief from his days in the hospital. "The operating room is completely controlled," he says. "The humidity is controlled; the temperature is controlled. You don't talk. It's sterile. You're sterile. And then suddenly you enter an environment where you get bloodied. You get cold. You get muddy. You get snowed on. It was the complete antithesis of the operating room, and I loved it." Once he retired, he started going out for several hours every day during pygmy owl breeding season to find pairs and their nests and to document their breeding behavior. Now he's one of the foremost experts in the United States on the breeding biology of this elusive little species.

"Steve knows more about Northern Pygmy Owl courtship than anybody in the country," says Holt. "He just busts his ass getting out there and doing it."

Hiro has found that during courtship, pygmy owls typically sing in the morning, but sometimes they'll sing all day long, calling to one another with their mating trill.

Courtship in owls is like this. No strutting around or flashing of splashy, colorful feathers, mostly just mutual hooting. "If you're an owl, you've got to sing to attract a mate," says Holt. "Big owls hoot. Little owls toot. That's their whole thing, vocalizations." But what hooting it can be, the choicest and most persuasive owl language.

Think of standing alone in a cold, dark forest. Suddenly, a deep, throaty *hoo-oo-oo-oo* of a Great Gray Owl booms through the trees and reverberates in the chest. A male Great Gray will often start seeking mates in late winter by calling through the afternoon and evening air, dazzling the female not with warbling song but with repetitive low-frequency notes that bid her come—*vvvvuh . . . vvvuh . . . vvvuh . . .*

vvvuh—like stones dropping through the air. The vocal courtship antics of Barred Owls during the nighttime hours in some suburban neighborhoods can be so full-throated and maniacal they banish sleep. Thoreau thought the courtship song of the Eastern Screech Owl sounded like the mutual consolations of suicidal lovers: "Oh-o-o-o that I had never been bor-or-or-or-orn" from one side of the pond, and a tremulous "bor-or-orn" echoed from the other. Male and female Blakiston's Fish Owls sing duets so seamless they sound like the song of a single bird.

The Northern Saw-whet Owl takes vocal courtship to an extreme. During the mating season, males fly rapidly around their territories, which are replete with a choice of nesting sites, tooting to draw in females. A male will perch in a safe spot, back to a tree trunk, and call continuously through the night to advertise himself, 112 toots per minute, from a half hour after sunset to a half hour before sunrise. (David Johnson has timed them.) By morning, he's hoarse. When a female comes into his territory, he'll ratchet up the speed of the call to 260 toots per minute. Then he'll show her his nest sites and give her a mouse to prove he's a good provider. If his nest sites are up to snuff and his food offerings satisfy, she'll stay with him. If not, she'll fly off, and he'll follow her, emphatically hooting at 160 toots per minute. When this fails to draw her back, he goes back to his perch site and starts all over again. Dominant or "winning" males need toot for only a few nights to find a mate. The "losers" must call and call, sometimes for weeks on end.

Once a male owl gains the interest of a female, he may start showing off in other ways. He might display his feathers by fluffing them out. He might stretch his head to the full length of his neck, then swing it to one side and drop it at least as low as his feet, then swing to the other side and raise it again. The details vary from species to species. Ornithologist Edward Howe Forbush described the "grotesque love-making" of Barred Owls as "ludicrous in the extreme. Perched in rather low branches . . . they nodded and bowed with half-spread wings and wobbled and twisted their heads from side to side, meanwhile uttering the most weird and uncouth sounds imaginable." Barn owls, too, can sometimes engage in curiously vehement courtship. Not long ago, Motti Charter, a researcher at the Shamir Research Institute, captured on video the strange and apparently hostile mating rituals of two barn owls in a big nest box. In the film, there's loud screeching and hissing in the background, and it's not at all clear whether the birds are mating or brawling. "They continue to struggle and jump on each other and do all sorts of other weird things," says Charter. "Initially, I thought it was a male that entered the box, and the two males were fighting, but it turned out it was a female. So," he says, "a very interesting mating ritual."

Contrast this with the soft love notes and snuggling of Burrowing Owls, sitting as close together as possible on their burrow, caressing each other with bills and head rubbing.

Like Northern Saw-whets, Great Gray Owls often woo with food. A male will give his advertising call around a nest structure. "He didn't make that structure, of course," explains owl researcher Jim Duncan, "because owls don't make their own nest structures, but he's hoping to attract a female to come and check him out and check out the real estate." The female listens for a call that interests her, and if she hears it, she'll work her way through the forest and land on a branch near him. "And then," says Duncan, "she'll look at him and go, 'Hey, you're kind of a good-looking male Great Gray Owl. And you've got a nice nest that you didn't make, but can you hunt?' She wants to check out how good a provider he is, so she challenges him to provide food by uttering a soft nasal *whoop* call." All winter long, these owls have been competing for food. But at some point early in the breeding season, something inside the male shifts. Instead of swallowing that captured vole, he'll carry it to a nearby perch and sit quietly. The female, in turn, will respond to the vole dangling from his bill not with aggression but by chirping for food. "She'll give what used to be called a 'food begging call' in the seventies," says Duncan. "Then a female biologist said, 'Wait a minute, female Great Grays are bigger than males. They're not begging for food. They're *demanding* it. So now it's more appropriately called a 'food demand call.' He'll go off and catch a vole and then fly in with it. And when he comes back with the vole, he gives a kind of accelerated version of his courting call that says, 'Look at what I got, how do you like me now?'"

When male Long-eared Owls display food to a female, they'll raise both wings so the two wing tips form what looks like a beautiful lyre shape. Duncan has caught on camera a similar pose from a female who's about to receive prey from the male. The female then turns her head sideways to receive the food. "It's kind of like when humans kiss," says Duncan. "We have this big schnoz we've got to move out of the way."

Hooting, fluffing, mutual preening, feeding. This is mostly what goes on in the pairing of owls. But one species takes courtship to great heights. The Short-eared Owl may have a call like an asthmatic dog, as naturalist Mark Cocker describes it, but its aerial feats of wooing will take your breath away.

Just before sunset on the open grasslands at the foot of the snow-capped Mission Mountains. It's mid-April and nearly everything is settling in for spring here, pelicans, pipits, Savannah Sparrows, Ospreys, harriers. A curlew calls from a distance and three Sandhill Cranes soar overhead, pink bellied in the setting sun.

I'm out to spot a sky dancer, the Short-eared Owl. As with so many owl encounters, you may hear the bird before you see him, his *toot-toot-toot-toot* fifteen, maybe twenty times in a row. Look up and you can spot him floating at great elevation. The Short-eared is the only owl species that does this—goes up, up in the air in daylight like a Red-tailed Hawk—and it does so only in a brief window at the end of the day. How high does it go? Two hundred fifty, three hundred feet? A member of the ORI team, Chloe Hernandez, sends up a drone to find out, but the device hovers at two hundred feet or so, well below the owl's height. The bird flaps his wings slowly and floppily, erratic as a big night moth. Then, suddenly, he takes a short sideways dive, followed by a dramatic upward swoop. A few minutes later, he does it again. Listen closely, and that slanted dive has a sound, like the flutter of a little flag in strong wind. As the owl drops, he brings together his long wings beneath him and beats them together with short clapping strokes, eight, ten, eleven times in rapid succession, as if he were applauding his own show. Then he flies up again and hangs in the wind as the world goes silvery pink and gray. He may repeat this dance several times to impress a female or for territorial display to rival males. Between displays males may grapple briefly in the air or just above the ground, tussling over territory or potential mates. In any case, it's quite a performance.

The closely related Long-eared Owl also sometimes flies high in this mothlike flight to defend its territory, but rarely during daylight hours. Other owls may do a sort of joint courtship flight at night. David Johnson recalls one night in early spring in northern Minnesota years ago, when he was walking along a frozen lake under a three-quarter moon. "A pair of Barred Owls were calling back and forth," he says. "Then, suddenly, I saw them flying in tandem over the ice, swerving together, lit by the moon. It was magic." We hear the caterwauling and conversation of courting owls, says Johnson; we rarely see the courtship flight, though it happens all the time.

But to see the Short-eared Owl displaying openly like that, high in

Short-eared Owl

the air in daylight: it makes the owl seem so—well—*vulnerable*, stalling, stooping, clapping its wings beneath it.

All this to impress a female. What dictates her choice of males? How does one owl choose another as its mate? It's still largely a mystery, but in some species, fascinating clues and theories are emerging, including one that explains the plumage of Snowy Owls, a strange departure in the owl kingdom.

The Snowy Owl is the only owl species with distinct sexual dimorphism in its adult coloring and plumage patterning. In most owls, you can't tell the sexes apart by their feathers. Adult male Snowies are a pure, almost fluorescent white, which makes sense for camouflage in a bird that lives most of its life in the high Arctic, where in summer there's close to twenty-four hours of daylight. Females are darker, with dark spots and brown bars on a background of white. "That's consistent with all open country species of owls, where females have more plumage markings than males," says Holt. "They tend to be overall a little bit darker, maybe for enhanced camouflage during the nesting season." But in Snowy Owls, females are dramatically darker than the pure white males.

Snowy Owls are thought to share a common ancestor with Great Horned Owls and from which it diverged some four million years ago. As Snowy Owls evolved and expanded into Arctic habitat, natural selection favored the evolution of white plumage. But Holt argues that mate choice and sexual selection have also been drivers for the male's pure white plumage. "Female Snowy Owls choose to breed only with males that are really white, almost fluorescent white," he says. "Of two hundred eighty-five nests that we found, every single nest had only fluorescent white males breeding. It takes a male three or four years to develop this coloration. Younger males look more like females in plumage, with the same brown markings, and they don't have territories. They sit on the sidelines and don't breed." So a male's plumage is a way for a female to assess the relative age, social status, and genotypic quality of a potential mate. "She can say, 'Hey, I've got an adult,' number one," says Holt, "and number two, 'He's got some resources.'" (Bright plumage can be an "honest signal" in the bird world, indicating high-quality genes and a good diet.) "'My chances of raising a family are better with him than with the young boy over here who may be great in the future but right now doesn't have anything.'"

Screech owls and Burrowing Owls seem to have what's called "assortative mating," when an individual picks as its mate a bird like itself in important ways. When David Johnson measured the body size and wing area of seventy-five mated pairs of Burrowing Owls in Oregon, he found that there was a positive linear relationship in the wing size and weight of males and females—that is, large females picked large males for their mates, and small females picked small males. "That seems unusual," he says. "We're so used to the idea that females want to mate with the big guys. But that's not the case with Burrowing

Owls. And as we're finding out, it's not the case with other owl species as well."

Why would owls pick mates their own size?

It has to do with energetics, says Johnson. "It's only advantageous to be big during a good food year. If it's not a good food year, it's better to be smaller because you need less energy to succeed." Assortative mating guarantees variability in the population, which is important if food sources are unpredictable. "It ensures that some birds in the population will be successful, and not everyone will crash and burn. So there's a real evolutionary pressure for owls to be small. That's something we didn't know or expect."

And there's more to it. Females also tend to pick males their own age. How can they tell the age of a prospective mate? "Through obvious signs males put out about their status," says Johnson—the depth and pitch of their vocalizations, their hunting prowess, their nesting success. "Owls have had millions of years to work this out. We underestimate their capabilities. They're a lot cleverer than we give them credit for."

HOUSE HUNTING

Owls may be skilled hunters and mate finders, but skilled nest builders they are not. Snowy Owls and Short-eared Owls come closest to making their own nests, scraping out a simple circular depression on the ground. Short-eared Owls tend to simply nestle in the grass, scratching out a bowl-shaped pit and filling it with grass and downy feathers. Snowy Owls excavate their shallow nest bowls on mounds on the Arctic tundra. In a long-term study of 280 Snowy nests, Denver Holt found that they cluster in hot spots with certain geographic characteristics: taller mounds on ridges that offer higher, drier ground for chicks, more exposure to a breeze for cooling and relief from mosquitoes, and perhaps most important, a panoramic view of the surrounding landscape and potential predators.

Most owls don't construct their own nests at all, but rather they appropriate structures built by other animals. The male usually finds a territory with abundant prey and some good nesting possibilities, but the female selects the actual nest sites.

She knows a good site when she sees it. Some species, like Long-eared Owls and Great Horned Owls, settle into old, abandoned stick nests vacated by magpies, crows, or hawks. Barn owls typically nest in the rafters of barns, in empty buildings or silos like that owl family in Linville, or in cavities along cliffs. Burrowing Owls live up to their name by nesting in underground tunnels dug by ground squirrels, prairie dogs, badgers, or other burrowing animals. The smaller species—Northern Pygmy, Northern Saw-whets, Flammulated, screech owls—find natural tree cavities or ready-made holes created by woodpeckers, sometimes in snags, standing dead trees that may be missing their top and most of their branches.

Taking up residence in a prebuilt nest is not a bad strategy. It means a breeding bird doesn't have to expend energy to build a nest of its own. But often there's an overall scarcity of good real estate as well as competition from other nest borrowers such as red squirrels.

The co-opting of other animals' structures is in part what makes owl nests so hard for researchers to find. "No question that locating nests is difficult and time consuming," says Holt, "but it's essential to understanding what's influencing their breeding success, what's going on with their populations—and if they're declining, finding possible ways to respond through good management." During one recent breeding season, Holt's team found nine Short-eared Owl nests. All except one failed. "It's crucial to figure out what's causing nest failure so we know what steps to take to protect the species."

Once the teams find a nest, they record its location, note its characteristics, and sometimes band the birds, both adults and chicks. They work quickly to minimize harassment, handling the birds as little as possible.

The nest search begins with locating the birds. Early in the breeding season, when owls are establishing or defending their territory, Holt's team sets out into the field with a Foxpro game caller loaded with owl calls. "When males are in territorial mode, it's easier to get a response from some species," says Holt. "Males take it as a challenge. 'Who just came into my area?' He's concerned that a new call might represent another male intruding on his territory. He'll usually fly to a higher perch and call for a few minutes. If he doesn't hear the intrusive hoot again, he'll leave. Once mates are found and nesting begins, all owl species quiet down, so it's important to locate them at this early stage. You can't find these birds if you show up casually," says Holt. "You have to be methodical, and you have to be out there. A lot."

There's no cutting corners with owls, as ORI researchers Beth Mendelsohn and Chloe Hernandez will tell you. The two often bundle up on freezing wintry nights and head out into the deep woods to survey for Great Gray Owls in remote areas on the western side of the Mission Mountains, Grizzly country. They conduct point surveys of the forests in the pitch black, sometimes on foot in the snow, sometimes on an ATV, pausing periodically to play on the Foxpro the booming territorial call of the Great Gray. They work from a half hour after sunset to 1:00 or 2:00 a.m., stopping every quarter mile or so to play the owl calls, listening for ten minutes, and then moving on.

Night work in bear country: searching for owls is not for the faint of heart. "It's not uncommon for bears to be found around Great Gray nests," says Mendelsohn. "They prefer the same types of habitat and sometimes even eat the eggs and chicks." But, she says, she has developed some comfort with it. "The more time you spend off trail in bear country, the more you get a feel for it. Instead of giving in to fear, I try to develop more awareness. When you're searching for owls, you have to be quiet because you rely on hearing them hoot. So we're trying not to make a lot of noise, but all of our senses are on alert and we're pretty aware of our surroundings.

"When you finally do hear an owl," she says, "it's like you almost feel it. It's such a rare experience, even for those of us who are out there so many nights. The low, deep hoot starts far off, so soft you're not even quite sure if you're hearing it. You get very still and try to be even more quiet. Then eventually you realize, yes, it's a Great Gray, maybe even a pair. It takes your breath away."

Once the birds are detected in an area, whether it's in the grasslands of the valleys or the deep mountainous woods, what follows is day after day of nest searching, combing fields and scouring forests for signs of owl occupancy.

Short-eared Owls may nest in open grasslands, but their small, scraped nests are not easy to find, well concealed in the tall grasses and forbs. "Plus, the female's main line of defense is to hold perfectly still until she is practically stepped on, relying on camouflage as potential predators walk right on by," says Mendelsohn. So in May and early June, Holt's team "drags" the fields with 200-foot sections of climbing rope tied together and attached at each end to an ATV or hand held by a hiker. The drivers or walkers move in parallel slowly, with just enough tension in the rope that it rides high in the vegetation so there's no risk of breaking eggs. When an owl flushes, the dragging halts, and the team searches for eggs or evidence of a nest, which they mark and record with GPS.

It's hard work. The fields are rough and weedy, and the ropes often won't go through the thick patches of teasel, an invasive plant that looks like thistle with short, tough nubs. And of course it's possible to miss nests. Twice the search team has seen a female Short-eared Owl jump over the passing rope and return instantly to her eggs. Only diligent watching revealed the nest itself.

Not all owl nests are so hard to locate. I think of the Burrowing Owls at the army depot in Oregon and all over the Americas, which ornament their nesting burrows with so many elaborate decorations they act like billboards. David Johnson has found burrows bestrewn with cornstalks, corncobs, the vertebrae of mule deer, moss, chunks of grass (at one nest he found 105 pieces of grass root wads), swatches of fabric (they prefer red, white, blue, and green, in that order), cattle dung, bison dung, Pronghorn dung, coyote scat, pieces of concrete, old gloves, even dehydrated seed potato pieces—a slice of the potato that has an eye in it. One season a farmer near the depot had finished planting his seed potatoes, and he dumped a truckload of what was left about a mile and half from the depot. "The male owls congregated all around it," says Johnson, "and everybody took home seed potatoes to decorate, carrying them all that distance. The females must have been like, 'Come on, Bob, really? Seed potatoes?' Are they really that hard up for decoration?" Johnson muses. "Or do they just want to make a statement? Because they certainly put a lot of energy into it."

Johnson has figured out that all this bedecking is not about mate attraction or courting. The male starts ornamenting only after the female has begun laying eggs. It's partly to provide soft nesting material for her to shred and line the nest cup, but it goes well beyond that. You can't really shred a corncob or a seed potato. One clue: males are very specific about what they bring in and where they put it, and they keep track, Johnson says. One day he experimented by putting some random materials at a nest site. Then he went to lunch for an hour. "And when I returned, this guy was calling loudly, territorial calling. He knew that somebody else had put decorations in front of his burrow, and he wasn't having it. He's bringing in goods for the female to line the nest cup, but he's also announcing dominance to other males and advertising to future females: 'I'm a tough guy. Look at all this stuff I collected.' If you want to show you're a tough guy in the world of Burrowing Owls, decorate! Decorate with chunks of concrete. Nothing says tough guy like concrete decorations. Males 'overprovide' to convey to other males that this place is taken."

It's a big flag to their species—and to ours—that the burrow is occupied. One male decorated his burrow with 122 pieces of coyote scat. "He had an airstrip going there," says Johnson, "just a strip down the middle left open for the landing. It was decked out on both sides, inches deep in coyote scat."

Hard to miss that.

The cavity nests of Northern Pygmy Owls? Not so obvious.

At a beautiful stand of aspens north of Polson, Montana, not far from Flathead Lake, Denver Holt and Jon Barlow, an ORI volunteer, show me the search strategy for cavity nests. Holt calls the stand Tribal Aspen Grove because it lives on land belonging to the Confederated Salish and Kootenai tribes. Each of us picks out a "whacker" from the back of the rig, a solid hickory stick about eighteen inches long. At first, I think it's to ward off the Grizzlies that frequent these woods. But it has another use.

We work through the dense undergrowth of willows and hawthorns, looking for aspens with telltale holes in their trunks, the cavi-

Northern Pygmy Owl peeping out of a nesting cavity

ties made by woodpeckers. Several trees—most, in fact—have the wounds of bear tracks on their white bark, and most have cavities. But not all cavities are created equal. We're looking for round woodpecker holes. The rectangular holes made by Pileated Woodpeckers for feeding are not suitable for nests. Saw-whets tend to choose big cavities with large openings in Ponderosa, aspen, larch, or cottonwood trees, excavated by Pileateds or Northern Flickers. Northern Pygmy Owls go for much smaller holes—natural cavities or holes excavated by sapsuckers or Downy Woodpeckers, anywhere from five to seventy feet in elevation. The cavities are often hard to see, cast in shadow. "You have to know what you're looking for," Holt tells me.

"A beauty here!" he calls out. "Cavity facing me. Give it a whack, Jon."

Barlow makes his way to the tree and whacks it hard two or three times with his stick, then rubs the stick up and down the bark, mimicking the sound of a predator climbing the tree. Out of the hole pops a red squirrel. If a female saw-whet had been nesting in the cavity, she, too, would have popped her head out jack-in-the-box style to assess danger from predators.

Banging trees in this way often brings out saw-whets, but not pygmies. Saw-whets will look out 90 percent of the time, says Holt, "but the cavities used by pygmy owls are so small, no predator can get in there, so it makes more sense for them to just hunker down."

There's nothing straightforward about the cavity nest search—for humans or for pygmy owls. The owls seem to have their own stringent and mysterious standards for what makes a good nesting cavity. They're picky, and they take their time making their choice. "I'll follow a pair," says Hiro, "and they'll seem to start concentrating on a cavity. I see them interact around it first thing in the morning. I see them copulate. I see the female go into the cavity for extended periods of time. So I'm convinced, 'This is it! This is going to be the cavity!' This may go on for two days. I relax. I figure we've got it, and now I can go on and focus on finding other nests. Then I come back to the site in three or four days, expecting to hear that morning vocalization around the nest, and I don't hear anything. I look in the cavity. Nothing. The pair has either moved on and found another cavity or they've disbanded. The owls had not yet decided which cavity they would choose. They were just house hunting!"

What was good about the cavity that kept them there for two days? Hiro wonders. And what was bad about it that made them move on? "I'm not sure we'll ever figure that out," he says. It's not about inadequate food supply. In one case, the pair moved only 200 yards away and then ultimately nested successfully.

Dave Oleyar, who studies small cavity-nesting owls in the forests of Utah and southeastern Arizona, has surveyed and mapped potential owl-nesting cavities in his region and established a few basic criteria for an owl that's house hunting. "The cavity has to have an entrance at least three centimeters in diameter," he says. (That's for a tiny Elf Owl.) "It can't just be a little cup or bowl. It has to go down and have some depth to it." He scores the cavities on a scale of one to five, like a real estate agent grading houses on the market. "There are a lot of cruddy cavities, to be quite honest, but there are also good ones. A high-quality penthouse-like cavity where I feel like I could crawl in there and the floor's nice and open? That would get a 'five.' Versus one where I can barely get in it and the floor is really jaggedy. A 'one.'"

The Tribal Aspen Grove is chock-full of what look like great cavities, at least from the ground, but no nesting owls. The search is like this, says Holt. "You whack a tree, a saw-whet pops out, you've found a nest right away, and you think, 'Okay, here we go.' And then you whack nine hundred more trees and get nothing.

"This is why researchers do nest box studies on owls," he says. "It's easier." There's no question that using nest boxes to study cavity-nesting birds has made it easier to monitor and capture both parents and their young, contributing to studies on the ecology and behavior of the birds. But Holt questions whether the data from these studies is reliable. "The placement of nest boxes rarely mimics natural conditions," he says. "They're conveniently situated along roads. They're spaced out and all one size." Natural nest sites tend to cluster deeper in the woods and to vary in size. "Also, pine martens and other predators remember the locations of the nest boxes, and they become like feeding trays.

"If this kind of work was easy, everyone would do it." Not everyone does. Very few, in fact. Which is one reason the work of the ORI—painstaking, meticulous, conducted over the long haul, season after season, year after year—is so important.

It can literally take years to discover the nests of Great Gray Owls, says Beth Mendelsohn. This is in part because of the owls' elusive behavior and camouflaged sites and the yearly fluctuations in their nesting, including years when they may not nest at all. "Looking for a Great Gray Owl nest is truly like looking for a needle in a haystack," she says, "hours and hours of climbing slopes, slogging through bogs and thick brush, scanning every tree for signs of nesting."

Discovering a nest, finally, can be oddly serendipitous and deeply joyful.

On Flathead Lake, north of Charlo, is a steep mountainous forest on tribal land known as Elmo, acres and acres of Douglas firs, big Ponderosa Pines, and larches that flame yellow in the fall. It's good Great Gray Owl habitat. While cavity nesters like Northern Pygmy Owls need natural tree holes or the drilled holes of woodpeckers, Great Grays rely on large broken-topped snags or the abandoned nests of other big birds. Larches spread their branches in radial wheels that make firm platforms for the goshawk nests that Great Grays sometimes adopt. The big owls also use the old stick nests of ravens or Red-tailed Hawks, sometimes hidden in mistletoe, a parasitic flowering plant that lives on Douglas firs and creates a dense, broomy thicket of erratic growth. These may not be preferred sites, Holt thinks—the owls favor snags—but are sometimes used by necessity.

Mendelsohn and Solai Le Fay, an ORI intern, recently heard a Great Gray calling in this area—a female, they think—making a kind of *gwuk* contact call in response to the Foxpro, and now we're looking for her nest. Holt and I spend a morning scrambling up and down the steep mountainside with the young researchers, looking for signs of the nest—old snags that might harbor owl nests on their broken tops, pellets on the ground, feathers indicative of a brooding bird. Right before a female is ready to incubate her eggs, a surge of hormones "de-feathers" the brood patch on her belly so she can press her warm, bare skin against her eggs, transferring more direct body heat to keep them at 100 degrees Fahrenheit or so. An increase in white blood cells wrinkles the skin across the brood patch. What would be the evolutionary reason for the wrinkling? Holt wonders if it prevents the avian equivalent of bedsores when a female is brooding, like those foam waffle pads used for hospital patients. In any case, on the ground or caught in branches, stray downy feathers from a brooding female often flag a nearby nest.

It's a game of looking up and looking down, and I think of what naturalist Scott Weidensaul once said to me: "If you think 'warbler neck' is bad, sister, try 'owl neck'—the crick you get from looking straight up into eighty-foot pine trees trying to find an owl." I also think of Max, the owl-detective blue heeler, and yearn for his knowledgeable nose. In places the undergrowth is thick with willows and alders, almost impenetrable. There are plenty of "whips," as foresters call them, downed trees, some leaning against the towering larches and pines. These are important for nesting. When the owlets fall out of the nests, which they often do when they're branching—moving from the nest out to branches before they fledge—they'll use these downed trunks and branches to climb back up the tree.

This is beautiful Great Gray territory. On the forest floor are small shrubs of snowberry and Mountain Spray, bright patches of spring beauty and Sagebrush Buttercup. Lichens known as Old-Man's Beard droop from the pines. Sprigs of Wolf Lichen spring from the Douglas firs, a gorgeous, almost iridescent lime green. Wolf Lichen is rich in toxic vulpinic acid and in the old days was boiled up with meat and used to poison wolves. Though it's dry terrain and hunting might be hard here, there are good nesting sites, cool and shady, with some impressive snags rising thirty or forty feet, with bowls large enough to accommodate the massive belly of a brooding female Great Gray. Le Fay circles them to spot feathers or pellets.

Seeing snags this way, through the eyes of owls, changes the way I think of these trees, still standing but now dead or dying. It's easy to love a living tree, with its lush foliage and canopy of greenery. But snags are like skeletons. They've lost their leaves, sloughed their skin. Their bones are furrowed with insect tracks, riddled with holes, rotted at the core, and their tops are stunted and snaggled. But what life they support! More than a hundred species of birds, mammals, reptiles, and amphibians use snags for nesting, roosting, denning, and feeding, including these magnificent owls. Now when I see snags made into roadside guardrails and benches, displayed as hotel "totem poles," or cut into cords and stacked neatly as firewood, I wonder at the loss.

A day or two later, bingo. Mendelsohn, Hernandez, Le Fay, and another volunteer had come back to comb the Elmo forest again, some 200 yards from where we had been looking earlier—an area the team had skirted in their search the year before because of a big black bear on the slope, sitting on his haunches in the sun. "I remembered some old snags there and thought they'd be worth checking again," says Mendelsohn. "As I scrambled up that steep hill, the crew was waiting at the bottom, and they heard two *whoops*." It was the female Great Gray. "I didn't hear it over the sound of my huffing and puffing. But then she vocalized again a few times, doing a single or double hoot so soft we were barely even sure we heard it, and we were trying to figure out which direction it came from. An hour or so later, having homed in on the sound, I looked up and saw one corner of what I believed might be a nest way, way up in this Ponderosa Pine, and I thought, 'Oh my God, that's it!' Sometimes when you see it, whether or not you can actually see the owl, you know that just has to be the nest."

A raven helped them zone in on it. When the bird flew by cawing, the female owl reacted with her *whoo-whoo* defensive hoot. With a camera, the team zoomed in on the top branches of the pine, and there ninety feet up in a clump of sticks—an old Red-tailed Hawk nest—they could see, just partially, one tail feather of a Great Gray Owl. The nest!

"I would bet that was also the nest site last year," says Mendelsohn, "and we missed it because . . . well . . . bear."

That night, the team was aglow with the finding, telling and retelling the moment of discovery. But Mendelsohn was also deeply concerned. "Most of these forests are managed for timber and hazard reduction, not for wildlife—and if the latter, mostly for elk and deer," she told me. "We work with forest managers and timber harvest projects, even responsible foresters who think they're never going to cut down a tree with an owl nest. But it took us two years to find that nest even knowing one was there somewhere—it was just so hard to see it from the ground. It's really scary to think that that tree could so easily have been cut."

This is why it's so urgent to locate and map owl nests, whether in cavities in an aspen grove or on broken-top snags or in old hawk nests high up in a Ponderosa Pine. If the nests aren't found and flagged, their host trees may come down.

If it's hard for humans to find owl nests, it's hard for owls to find them, too—and getting harder.

Think of what it takes to make a cavity or hollow suitable for an owl to raise a family—say, a Northern Pygmy Owl's nest in an aspen tree. It takes a climate with just the right conditions for aspens to grow and thrive for a long enough time that some will get injured or begin to die. For a cavity to develop in a dead or injured tree, it takes fungi that weaken the wood and make it softer and therefore easier for excavators like woodpeckers to drill a hole. The process can take many decades.

The same is true for the large hollows or snags with big broken-top bowls that are acceptable sites for big owls like the Great Gray.

Throw in human activity, the removal of big timber, the culling of dead trees, the harvesting of snags for firewood, as well as wildfires, lightning strikes, and other natural disasters, and it all seems fairly tenuous. This is true not just in North America but all over the world.

In Australia, as in North America, most owls depend on hollows or cavities that are often few and far between. "All Australian owls except the grass owl are obligate hollow nesters and don't take well to nest boxes," says Steve Debus. "Boobooks will occasionally use a nest box and barn owls sometimes use an artificial structure, but basically, Australian owls all depend on hollows." As big, old trees fall to timber harvesting, wind events, and fire, so go the big hollows. What remains are smaller trees with smaller hollows.

"Some big owls, like Powerful Owls, literally have to hitch their shoulders sideways to get in," says Beth Mott, a conservation biologist studying tree hollows and the fauna that use them. The rarity of hollows spurs aggressive competition. Mott has observed Sulphur-crested Cockatoos removing Powerful Owl chicks from hollows. "They're really the bullies in this story," she says. So are Australian Magpies and currawongs, which attack chicks and sometimes kill them. "This direct aggression by birds who want to use the hollow themselves, resulting in the death of chicks, is something that we're seeing far more of as time goes on."

Mott thinks that hollows are a unique environment, not easily replicated in a nest box. For one thing, hollows that form in living trees buffer against temperature and humidity changes in the environment. Nest boxes tend to be hotter than tree hollows, "which when you're a mother bird sitting on an egg is a big deal," says Mott. "Particularly when you're a Powerful Owl, which is a bird species that's really prone to heat exhaustion." Also, she says, "the microclimate that exists in-

Powerful Owl chick in a nesting hollow

side a tree chamber is an incredibly dynamic recycling environment that we don't see when we try to replicate a hollow with something artificial. If you look in a nest hollow just after a family of Powerful Owls has left it, you don't see big piles of possum remains or fur. All that's left is what we Australians call 'mud guts,' the friable material inside the tree where the termites have been eating. You might see a few bones, sometimes a few backbones, but mostly it's just mud guts, and it looks incredibly clean. With that huge influx of food and no cleaning up—you know, they don't take the trash out—all of this recycling is going on inside the hollow by invertebrates, fungi, and bacteria. So I don't think we can replicate in a nest box that really special dynamic environment offered by a tree hollow until we really understand it."

Dave Oleyar and his graduate student Kassandra Townsend are also looking into the thermal properties of hollows in different tree species and thinking about the effects of global climate change on the cavities themselves and on the owls and other animals that depend on them. "Will the same suite of animals use these cavities when (or if) the temperature range within falls above or below certain thresholds?" Oleyar wonders. "We're also starting to think about the implications for the microbiota that occur in the cavities—will decay rates change? What will this mean for the rates of cavity creation and loss? And for the owls that rely on these hollows?"

Mott describes a nesting hollow in a tree that had been used for more than ten years by a pair of Powerful Owls. Fires ripped through the area and burned the tree from the inside out. After the fire, the pair sat in an adjacent tree. "Large forest owls make a really obvious grieving noise when they've lost a chick," she says. "And they did the same thing for this tree hollow. It was quite heart-wrenching."

Once these sites are gone, it can take decades or longer to replace them. Mott is working to develop strategies for protecting hollow-bearing trees in Australia and for retaining big trees that haven't yet developed hollows, "recruiting" for the next generation of tree hollows. Experts agree that we need a long-term conservation strategy to plan for and protect all the life stages of trees, from cavity-bearing live trees to snags and woody debris, so that there is a continuous supply of nesting habitat in forests around the globe.

Back in Montana, it's laborious, methodical searching that often yields a Great Gray Owl nest on a snag or in a stick nest. But sometimes it's chance.

Hidden in a forest in a low boggy area, a place dubbed Grizzly Hill—for good reason, the mud is full of bear tracks—the ORI team shows me the site where Mendelsohn recently made an unexpected find. She was scouting for a night survey with Jon Barlow, and they

Great Gray Owl chicks in the snag nest at Grizzly Hill

had paused at the spot to allow Jon to relieve himself. "We were just wandering down the hill, noting the Grizzly tracks," she says. "When we stopped for Jon, I looked to the right of the trail and there she was, right there, a Great Gray Owl sitting on top of a big broken-top cottonwood snag. This is not the first time we've found a nest when someone had to pee."

At first, I see nothing but a big snag. But when they point her out to me, I can just see her huge facial disk peeking over the bowl as she twists her head toward us and gives us that placid Great Gray look. It's one thing to see one of these spectacular owls on a distant fencepost, lean and hungry and ready to launch. It's quite another to see a big female like this so nearby, tucked into her nest like a queen settled on a rough-hewn barrel-shaped wooden raft. From below, the hollow of the snag looks small and jagged, hardly like the comfortable natural bowl that would make for good nesting. But Holt tells me that as long as she has room to put her eggs down and get her abdomen over them, she can make do. "I just hope no one takes this snag before those chicks hatch."

RAISING OWLETS

Steve Hiro is focused on the nest of a Northern Pygmy Owl three stories up in a big aspen tree north of Charlo. He's peering up at the nest cavity and bracing a ladder against an adjacent tree. At the top of the ladder, Denver Holt is wrestling with bungee cords to strap the shifting ladder to the tree while four of us look on nervously—Hiro, Le Fay, Hernandez, and Mendelsohn, who's especially anxious. Dangling from Holt's right arm is a camera. "Careful," says Mendelsohn. "That's a six-hundred-dollar camera—and it's not ours."

Holt is trying to position the camera so that it will take a steady stream of photos of the nest in the cavity directly across from it. The camera is motion triggered, so it should capture all kinds of activity, from the male delivering prey when the female is on the nest to the fledging of young. But Holt must situate it blindly, as he can't see around the tree. Mendelsohn has a spotting scope aimed at the camera to direct him.

Hiro has been waiting for this moment for years. The team tried putting a heat-triggered camera on pygmy owls at the Pattee Canyon nest, but the birds were too small to trigger the temperature sensor. Holt finally gets the camera properly positioned and descends the ladder. But it isn't working, so he climbs back up. "I'm old, but I still climb better than anyone," he says. He's fit, all right—the rigors of research on the Snowy Owl in the Arctic see to that—but when he's down again, we all breathe a sigh of relief.

"It's always more complicated than you think, a four-person job," he says, and adds with a mischievous grin, "If it was easy, anyone would do it."

Hiro spots the male in an aspen nearby and then the female in a nearby tree. "So she's not in the cavity after all. Maybe she's taking a break from the nest," he wonders aloud. "But if she's incubating, she should be in there." There's concern in his voice. The team decides to look inside the nest with a "peeper cam," a tiny camera attached to a fifty-foot telescoping pole that allows them to peer into otherwise inaccessible cavities. Not only is this cavity thirty-three feet up in the tree, but the opening is tiny—two, at most two and a half, fingers wide—typical of the kind of cavity selected by pygmy owls and probably drilled by a Hairy Woodpecker or maybe a Red-naped Sapsucker. Holt and Mendelsohn raise the peeper cam to a level where it has a good view inside the cavity. It takes some maneuvering, but what finally comes into focus is two beautiful round white eggs.

"Ah!" exclaims Hiro. "It makes sense why she's out. She's not done laying."

Here's one way that Northern Pygmy Owls are rebels in the owl world. The females in most species of owls start incubating as soon as they lay an egg. Not a female pygmy owl. She'll lay an egg every day or day and a half, with a total, usually, of five to seven eggs. But only after she has laid all of her eggs will she settle in to incubate. The eggs are kept viable by hormones while they await the warm brooding that will launch their development.

The number of eggs she lays, as with all owl species, relates to her condition, says Holt, and how much food is available, both where the owl is nesting and—in the case of migratory species—where she came from. "For example, if local vole populations are high, a female Short-eared Owl might lay as many as ten eggs," he says. "If the vole population is low, she might lay just three or four eggs, or none at all."

The whiteness of the eggs we see in that little cavity is usual for owl eggs but unusual in the bird world at large. Most bird eggs are elaborately colored for camouflage. I think of the speckled and streaked eggs of shorebirds that blend in with rocks, pebbles, and sand to bewilder egg thieves like snakes and foxes. Or the uniquely colored and patterned eggs of Japanese Quail, which the female matches with a nesting area that makes the most of their camouflage. Only birds that lay in places hidden from view have white eggs—woodpeckers, bee-eaters, kingfishers, owls. The eggs are concealed, so there's little reason to expend precious energy producing pigments for them.

Owl eggs are also different in shape from most bird eggs, rounder, probably for reasons having to do with flight, according to new research by Princeton University evolutionary biologist Mary Caswell Stoddard. The eggs of birds that are airborne most of their lives—murres, sandpipers, albatrosses—are elliptical, the theory goes, streamlined to fit through the pelvis of the birds' smaller, lighter, and more compact skeletons. Owls, which tend to fly only in short glides, have a heavier skeleton and a wider pelvis that can accommodate more spherical eggs. (But there are exceptions, of course. Barn owl eggs are decidedly conical or ovoid. The eggs of Great Grays are "egg-shaped," too, but they're smaller than a hen's egg and can fit through the bird's pelvis.)

While the round white eggs of the Northern Pygmy Owl are typical of most owl eggs, what's not typical is how they hatch. Look at sibling chicks in the nests of most owl species, and you'll marvel at the way they vary in size, sometimes so dramatically they appear to be different species (like the barn owls in that silo in Linville). Some chicks might already look like small adult owls, still downy but with the dark facial disk of adults and flight feathers already growing in, while others look like tiny balls of white downy fluff. Owls lay their eggs hours or days apart, and in most species, those eggs also hatch hours or days apart. It's called "asynchronous hatching," and it's the way most owls ensure the survival of at least some offspring. This makes sense, given the unpredictable nature of their prey populations—voles, lemmings, and other small mammals. If prey is abundant, all the young will likely survive. But if food becomes scarce, the parents can't feed all of the chicks. The first-hatched siblings—bigger, stronger, and better able to vigorously beg for food—will get most of the meat and have a chance of making it. The younger owlets may starve to death.

"It's a brutal world for the chick that happens to be the last to hatch," says Jim Duncan. ("I'm glad I'm the middle child in my family," he quips.) This may seem inefficient, even cruel, but it raises the chances that at least some nestlings will survive to healthy maturity. Staggering the hatching of chicks also limits the food demands at any given time because the nestlings vary in their stage of development. Finally, getting a chick out of the nest as fast as possible reduces the risk of a raiding predator taking the whole clutch.

But some species of owls are rule breakers, and—as Steve Hiro discovered—Northern Pygmy Owls just seem to do things differently.

When Hiro found a pygmy owl nest only seven feet off the ground in the snag of a larch tree, he knew he had a unique opportunity. "I realized that no one really knew anything about the hatching, development, and fledging of pygmy owl chicks, and I could document the whole process in detail with photographs. I didn't even need a ladder. I could just look into the nest with a borescope"—an optical instrument used to inspect the inside of a structure through a small hole.

Hiro started taking pictures on the second or third day after hatching and saw that the chicks were developing at exactly the same rate even though the eggs had been laid asynchronously. "The change in their down, the loss of egg tooth, their eyes opening, the flight feathers starting: you could see that at every phase along the way, they were developing at the same pace." By day twenty-four, the chicks had started looking out of the cavity and were getting increasingly bold and curious. Hiro wanted to make sure he didn't miss their fledging, so he went every day to check on them. On the morning of day twenty-eight, he arrived at five thirty in the morning, looked into the cavity, and saw five chicks. "So I went home, took a shower, and then came back at around 9:00 a.m.," he recalls. "And I said, 'Oh, just for the hell of it, I'll look in again.' There was only one chick left!" In just three hours, four of the chicks had left the cavity. With other species of owls, this would have happened over a period of days.

"I thought, 'Well, where the hell are they?'" says Hiro. "I looked on the ground and in the low branches around the tree. No sign of them. Then I looked up, and I saw these little things with no tail, way, way up in the tree. They were thirty feet up in the air, flying back and forth between branches thirty yards away. I couldn't believe it. Within hours of leaving the cavity, they could fly. And they were not just gliding between branches, but actually flapping, flying upward. It just seems amazing. How do they gain the strength to fly that way? How do they develop their wings when they're still inside the cavity?"

In any case, Hiro had photographically documented synchronous plumage development, hatching, and fledging in Northern Pygmy Owls for the first time. Since then, the team has confirmed the finding in six other nests.

It all seems very un-owllike. (The only other species known to hatch this way is the Little Owl, which sometimes hatches synchronously and sometimes asynchronously.) Why don't pygmy owls have the same food constraints as other owls? Why aren't they subject to the same pressures of predators?

The nuthatches and chickadees may have an answer, at least to the first question. Northern Pygmy Owls eat a lot of birds. This provides them with a relatively stable food supply compared with other owl species that depend on rodents, which tend to fluctuate in population. And the cavities they nest in are tiny, too small for most predators to access. Also, the ability to fly straight from the cavity without branch-

Two-week-old Northern Pygmy Owl chicks inside a nesting cavity, photographed by Daniel J. Cox / naturalexposures.com

ing means these little owls are not vulnerable to ground predators. But it remains a mystery how they exercise their wings before they fly.

Once that female pygmy owl is incubating, she'll leave the cavity only twice a day for five to ten minutes to regurgitate a pellet or to defecate. The male brings her meals as often as every two hours. But after the chicks hatch, demand for food increases exponentially, and by day ten or eleven, the chicks' needs are so great that the female joins the male in hunting.

In many ways, owls appear to be exceptionally dedicated parents. While a female incubates, she sits on the nest for nearly twenty-four hours a day, with her head low and stomach down, keeping the eggs at the right temperature, cool in the heat, warm through snow and cold.

A brooding Snowy Owl must hold the eggs at close to 98 degrees Fahrenheit, even when temperatures plummet to −30 degrees or −40 degrees. A Great Gray Owl nesting in the sun, on the other hand, may get very hot herself but will sit tight to keep her eggs and chicks from overheating, cooling off her own body through panting called "gular fluttering." I've seen a male Great Horned Owl spread his wings over the female and chicks to shield them from the pouring rain, drenching his own feathers in the process.

Jim Duncan recalls an especially stressful parenting experience for a pair of Long-eared Owls nesting in Manitoba one May day. "I remember being in the house and hearing this commotion to the north, where the nest was," he recalls. "And I looked out, and there were at least fifteen to twenty crows mobbing the owls, and I thought, 'Oh, no! If the female leaves the nest, a crow swooping down will easily grab the eggs.'" She stayed. "But then we had rain that day, which turned to snow as the temperature plummeted to −4 degrees Celsius, and there were sixty-to-seventy-kilometer-per-hour winds from the north driving the snow horizontally. Two of the eggs on the north side of the clutch froze. But afterward we went back to check things out, and four of the six eggs had hatched."

Thanks to good parenting, those young survived a murder of crows *and* snow.

Owls sometimes face a massacre of mosquitoes, too. In his studies of Great Grays in Canada, Duncan has witnessed the impact of an onslaught of the bloodsucking insects. "These birds choose to nest in really beautiful habitats," he says, wet meadows and bogs that are also great habitats for mosquitoes. The owls usually nest before the major mosquito hatch, "which is smart on their part." But some years they nest after the hatch, "challenging for them and for us. You feel like a human pincushion," says Duncan. It's worse for the owls. Adults and older chicks can perch high up in trees, away from the insects, but if the younger chicks and the adult female are still in the nest when the mosquitoes hatch, they have no escape. "We've had some years where very few chicks were actually fledged or survived the nestling period because the mosquitoes pretty much drained them of blood."

Once the young hatch, male owls must bring in enough prey to feed both the female and the growing young. (Unlike pygmy owls, in most owl species, the female remains on the nest to protect it.) Flammulated Owl and Elf Owl fathers bring their chicks invertebrates, including scorpions—but only after removing the venomous stingers. Dave Oleyar remembers being "footed" while handling an Elf Owl, "which you would think was no big deal because it's a forty-gram bird," he says. "But ten minutes later, after we let the bird go, my finger was blown up to the point of . . . well, quite large. And we think that's because it probably had some scorpion venom on either its bill or its talons. That bird didn't seem to be affected, but I sure was."

Holt has seen male Snowy Owls stockpiling food, bringing in dozens of the biggest, heaviest Brown Lemmings and piling them up around the nest. At one nest, there were eighty-six of the rodents. Snowies don't have to worry about spoilage because of the cold and because the chicks can swallow a lemming whole by the time they're two weeks old.

Eastern Screech Owls in Texas bring live blind snakes to their nestlings, not just for food but perhaps to keep their nests tidy and sanitary as well. The chicks will eat some of the snakes, but most of the tiny (and seriously bizarre) serpents will live alongside the owlets in the nest debris and eat the parasitic fly and other insect larvae in decomposing pellets, fecal matter, and uneaten prey. Scientists have found that nestlings with live-in blind snakes grow 50 percent faster and experience lower mortality than broods lacking serpentine company.

A female Great Gray goes the screech owl one better, eating the feces and pellets of the young and then, a few times a day, flying off to regurgitate the mass. Now that's a devoted mother.

Successfully raising young is such an energy-demanding task that the adults routinely lose as much as a third of their body weight during the nesting season. In pouring their resources into their offspring, they also forgo molting their feathers—a process that consumes significant energy. Duncan has compared the feathers of a male that raised young with those of a male whose eggs didn't hatch. He found that by the fall, the male with the failed nest had molted nearly all his wing and tail feathers, so he looked "all shiny new," says Duncan. "Meanwhile, the male that we recaptured after raising five young looked like a spent salmon, his feathers all brown and worn." This can take a serious toll as worn and damaged feathers affect a bird's flight abilities.

Owls, especially female owls, also expend energy fiercely protecting their nest. In some species, the difference in size between male and female owls—with the female generally larger than the male—may relate to the role a female plays in protecting her young. Females often stay put while the male hunts, so they're in charge of fending off intruders.

Many owl researchers have learned about serious nest defense the hard way, usually when they're trying to band chicks. There's a period of about a week when the female is most aggressive in her nest defense, after she has been nesting for a couple of months, laying and incubating the eggs, feeding and raising the chicks. The young are ready to leave the nest. They are also just the right size for researchers to band them. At that particular moment, the female has peak investment in her offspring, so she's red hot in her defense.

If you ask Jim Duncan about his best owl scar, he'll tell you, first, that it's hidden and, second, that it came from a female Great Gray Owl defending her young from banding efforts. In Manitoba, where Duncan lives, Great Grays weigh about two or three pounds. "Yeah, they're superlight," he says, "unless they hit you. We always have someone spotting, so they can warn the person climbing the tree that the owl's coming in for an attack. And then you hug the tree really tight, and you wave your arm at the last minute, and almost always the owl veers away. The main goal is to get in there, do what you need to do, and get out as quickly as possible again, to minimize the disturbance to the owls." While studying one nest site, he recalls, "this big female came barreling in. For whatever reason, we didn't detect her until the last minute, and she got me in the back of the head. It was like being hit with a two-by-four with nails stuck in it. One of the talons dug deep into the back of my skull, and the tip broke off in there. The blood was streaming down—head wounds bleed a lot—and I pulled out a piece of the talon. Whoa."

The ferocity of nest defense varies from owl to owl, even within a species. Duncan says he's only been hit by Great Grays two or three times in the thirty years he has been studying the birds. "The rest of them, they look menacing as they perch nearby, but they don't actually come in and hit you." In Europe, Great Grays have a reputation for being extremely aggressive. The dramatic behavioral difference may have to do with how long the birds have been exposed to humans, Duncan muses. "You might speculate that Europeans had settled in some of those areas much longer than in North America. So maybe owls have become more aggressive through selective pressure. I'm not sure. It's fun to speculate. Piles of theories are this high," he says, gesturing to indicate three feet. "Piles of proof are this high"—more like an inch.

Denver Holt's best owl scar is hidden, too. It's from a female Snowy

Owl, among the bird world's most aggressive nest defenders. When a female Snowy sees danger approaching from as far as a quarter mile off, she'll leave the nest "and go sit around on the ground pretending there's no nest there," says Holt. "But if you start toward the nest, she'll fly in and either she or the male may attack." Holt has been hit hard multiple times on the head, shoulders, neck. But the strike he remembers best was when he was lying belly down working to band chicks in a nest. The female swooped in and hit him so hard from behind with her talons that she punctured the tough canvas of his Carhartt pants and his long underwear, he says, "putting four holes in my butt. She got me good."

Perhaps not surprisingly, one of Holt's students discovered that the most aggressive pairs of Snowy Owls fledge the most young. They're best at defending against not just human banders but nest raiders like Arctic Foxes, Wolverines, even Polar Bears.

Ural Owls in Finland are so forceful in guarding their nests that some banders there wear hard helmets to protect themselves. When ornithologist Pertti Saurola started as head of the banding lab in Finland in 1974, he issued instructions to all banders to avoid wearing a helmet unless it was leather or covered with a thick foam pad or other soft material. This was in response to at least five cases where the females slammed into banders' helmets so hard that the birds died from the impact. "Through the decades I've used a fur hat, which prevents the penetration of claws only partly," says Saurola. "But isn't it okay to get some punishment for being a disturbing intruder? Nowadays I pull the collar of my leather jacket over the fur hat, and it helps."

Chicks will sometimes defend themselves by wing spreading, bill snapping, biting, kicking up their legs, and hissing—the latter most ingeniously performed by Burrowing Owl chicks. When cornered in their burrow by an intruder, the chicks will buzz exactly like the rattle of a rattlesnake. "Batesian acoustic mimicry," as it's called—when a harmless species imitates the sound of a dangerous one—is not unique in the animal world. It's a kind of evolutionary "ruse" that protects the mimic from would-be predators. Scientists recently found that Greater Mouse-eared Bats imitate the buzzing sound of noxious stinging hornets to ward off predatory barn owls and Tawny Owls. But the hair-raising rattle-hiss of a Burrowing Owl chick is remarkable by any measure and has set more than one researcher backing away from a nest burrow.

Beth Mendelsohn says she wishes owls did more to defend their nests, especially Short-eared Owls, which nest on the ground. "They're just so vulnerable. They'll swoop small predators, like skunks, but it just doesn't seem like enough." In the recent breeding season that saw the loss of so many Short-eared Owl clutches, it's likely the nests were raided by predators. "We're setting up remote cameras to figure out what the heck is eating all the eggs," says Mendelsohn. "Has nest predation gotten worse? And if so, why?" The team thinks development of the region has something to do with it. Populations of ravens—major nest predators—have swelled over the years as people have settled the area and planted trees that ravens can nest in near their houses. In some places around Charlo, there are as many as three raven nests within a mile radius of a Short-eared Owl nest.

Mendelsohn says that the ORI gets a lot of comments about what great parents owls are from people watching their nest cams. "It's true, there's a lot of intense parental care—the female sitting on the nest for extended periods of time, protecting it, feeding the chicks at her own expense. When we capture a female at the end of the nesting season, her body condition can be quite poor because everything she has has gone into the chicks," she says. "And it's true for males, too. They end up skinnier because they're feeding whatever they can to the chicks."

But at the same time, there are instances of what looks like parental abandonment or neglect. A week or two into a recent nesting season, one of three Great Horned Owl chicks fell out of a nest in Charlo and landed on the ground. The chick died of exposure or starvation. No effort was made by the parents to feed the chick on the ground or get it back up to the nest. They just continued to feed the two young that remained in the nest.

"When I see things like that, it reminds me that a lot of this parental behavior is programmed," says Mendelsohn. "The parents see the chicks in the nest, they respond and feed them, but it doesn't go beyond that." This makes sense from a genetics standpoint, she argues. "To carry on the population, an adult breeding owl has to produce only one offspring that survives to adulthood and breeds successfully. So in species that live for many years, like Great Horned Owls, it's probably better not to risk trying to protect young that likely won't survive." Whether or not a parent will feed a "downed" chick probably has to do with its age. If a chick is older, with a better chance of making it, the parents are more likely to continue to feed and care for it.

Sometimes owl parents stand by during apparent atrocities perpetrated by their own chicks and do nothing to interfere. In his video *Wings of Silence*, a visual exploration of nine species of owls in Australia, filmmaker John Young shows a chilling scene in a hollow containing a barn owl nest: one owlet killing and eating its smaller and weaker sibling. It's a gruesome sight, difficult to watch. But it's not uncommon among birds such as hawks and herons. A webcam on a nest box in California caught another barn owl family in the act. The nest cam was highly popular, attracting tens of millions of viewers, but the hosts turned it off when two of the four owlets died—it's not clear how—so people wouldn't see the dead owlets being consumed by their siblings.

Scientists have also seen instances of cannibalism among other owl species. During times of food scarcity, older and stronger siblings will kill and eat the younger and weaker ones. Nestling screech owls have been known to fight fiercely among themselves for food and to kill their smallest sibling. Australian researcher Raylene Cooke found juvenile Powerful Owl remains in eight out of nine pellets she dissected from one Powerful Owl nest.

Denver Holt says he has never seen siblicide in owls or the cannibalism reported in some species. But he did once see a Snowy Owl chick, the fifth born of five chicks, that was sick and weak and wouldn't feed. The mother bird held the failing chick under her breast feathers, but after a while it fell out and died, and then she fed it to the other chicks.

Nestlings certainly compete for food, sometimes hoarding it. Jim Duncan has camera footage showing a Long-eared Owl chick receiving a deer mouse and instantly hiding it from the other chicks, "then trying to scarf it down as fast as it could."

But it's not all *Lord of the Flies*. Some owlets seem to display a remarkable form of altruism. Not long ago, Pauline Ducouret and her colleagues at the University of Lausanne found that older nestling barn owls can be impressively generous toward their smaller, younger siblings, donating portions of their food to them on average twice per night—a display of altruism thought to be rare among nonhuman animals. "As expected, these donations are made by the oldest chicks in good physical condition, especially when food in the nest is abundant," says Ducouret. "The oldest member of the progeny will be the most altruistic if parents favor it by providing it most of the prey items." In some cases, the "altruistic" nestling will give its prey to the sibling that's begging most intensely. By feeding a particularly hungry sibling, the theory goes, the older chick gives it a better chance to survive and reproduce, thereby promoting transmission of their common genes. In other cases, the donor chick gives prey to the younger sibling that has preened it the most. So it's more like an exchange of services, a give-and-take. Either way, the practice may be advantageous to all, as passing responsibility for the distribution of food to the elder siblings gives the adults more time to hunt.

Quite possibly the riskiest stage in the making of an owlet comes when the young are ready to leave the nest: branching. Once nestlings can regulate their own temperature, but before they can fly, they may branch, clambering out of the nest and wandering out onto a limb or even jumping to branches of neighboring trees.

"We think branching developed so that even before the chicks fledge, they can spread out, away from the nest," says Mendelsohn. "Things have probably gotten a little cramped for them, and they're extremely vulnerable to predators there. If they're all stuffed in the nest and a predator comes, they'll probably all be eaten." They can hide better in the trees.

The strategy works. When the ORI team tries to band chicks, it's a feat to find them. "We may know there are five chicks, and four of us will scour the nest and find only a couple of them," says Mendelsohn. "They'll get as high as they can and stay very still. When you do finally see one, it's usually staring straight at you, wide eyed, frozen on its perch. Once the chick realizes you're a threat, it will start doing all different sorts of defensive displays, wing spreading, bill snapping, and hissing."

There are perils to branching. Sometimes the owlets end up hanging by a bough, clinging by one leg with their talons, and then either hoisting themselves up or falling to the ground. To climb back into the tree, they'll use a slanted "whip"—one of those downed trees leaning up again the nest tree—or go straight up a trunk if there are enough branches, crawling upward with their strong feet, hooking their bill like a parrot, biting into the bark for purchase, and flapping their wings

Barred Owl chicks branching

hard. Some owls even have vestigial wing claws that may help with the climbing.

Soon after branching, owlets start taking short flights from one tree to the next. "It's kind of comical, fun to watch," says Mendelsohn, "like watching a toddler learning to walk. They get going and then they can't stop and just crash into something or fall over or just completely miss their destination." Chloe Hernandez once observed a family of Great Horned Owls fledging near the ORI field station. "They would execute these labored flights around the field station," she says. "Then, from dusk to dawn, they would beg relentlessly for food from their parents." Hernandez watched the parents bring in big meals—American Coots, Ring-necked Pheasants, Mallards.

This is another motivation for branching. Owlets jump from the nest to the ground to be the first ones to a feeding. David Johnson once found four Great Gray owlets walking along the ground, 200 feet apart, to meet their father, who was bringing in food. Fledglings know to stay close to the nesting site for regular feedings and will sometimes harass parents when clamoring for food. Hungry young Boobook Owls can get belligerent with adults. Australian researchers have seen large female nestlings chasing smaller adult males, "aggressively seizing food and mantling over it [defending it with drooped wings], and adult males shying away or 'chittering' in apparent distress." No surprise that late in the post-fledging period in these owls, the adults often roost apart from their offspring.

At this stage, owl parents sometimes feed owlets to the point of groggy satiation, and chicks may eat until they fall into a deep sleep, sometimes lying on their stomachs crosswise over a branch, like a sloth. Like human babies, baby owls sleep a lot, and they spend more time than adults in rapid eye movement (REM) sleep, the type of sleep associated with vivid dreams. In 2022, scientists found that REM sleep in mice involves a kind of cognitive processing that might help to shape behavior when the mice are awake, such as avoiding owls and other birds of prey. Its presence in baby owls may likewise have to do with thoughts that occur during dreams—reenactments or reinforcements of skills. The owls at the International Owl Center vocalize in their sleep, says Karla Bloem, chicks and adults alike. "It's pretty hilarious when they wake themselves up."

Gradually, chicks start to fend for themselves. At first, they're clumsy and ungainly and don't respond in appropriate ways to predators or prey. Fledgling Powerful Owls "are incredibly curious, but they don't know anything about the world," Beth Mott told me. "So when you're trying to be a good biologist and keep your distance, they will actually follow you through the forest. And they're like, 'What are you?' Tipping their heads upside down and flapping their wings." They also carry things around. "They call it bark ferrying and it's a way of practicing how to hold and handle prey," she says. Young birds living in urban and suburban environments carry strange stuff, plastic,

Great Horned Owl chick testing its wings

tea towels, pairs of shorts. "They'll practice hunting skills and dismembering prey by raiding human clothing off a clothesline and tearing it to bits. I suppose they see something flapping in the wind on a clothesline at night, and they fly down and seize it. It takes them a while to stop being so dorky."

Mendelsohn has seen young owls practicing pouncing on things. "Sometimes there's not even anything there," she says. "Maybe there was an insect I didn't see? But the chicks will go and fake kill it. They're practicing the motions."

Eventually owlets learn to hunt efficiently and avoid predators. How long the process takes depends on the species. Great Horned Owl youngsters spend six months with their parents, and if displaced from them will starve. Barred Owls stay "home" a mere six weeks and then are off on their own.

Egg to independent owl in weeks. It's astonishing.

Six

To Stay or to Go?

ROOSTING AND MIGRATING

Imagine being an owlet. You've had maybe six months with your parents, maybe as little as six weeks. You barely know the difference between friend and foe, predator and prey. Now you're on your own and faced with choices that will determine your survival. Chief among them, whether to stay where you hatched, the only place you know, and roost there, perhaps through a winter of cold, snow, and scarcity of prey, or move on to a new spot, which may be warmer with more abundant prey, but utterly unknown and perhaps at the end of a long and harrowing journey.

What's entailed in the staying and the going? Three species are showing us just how complex the calculus can be and how unusual the routes that some owls choose: the Long-eared Owl, the Northern Saw-whet Owl, and the Snowy Owl. All three birds break the owl mold in a most spectacular way, pointing to the complexities of the "decisions" owls make and calling into question what we thought we knew.

TO STAY . . .

Just outside of Missoula, Montana, Chloe Hernandez and I are winding our way through a gully thick with vines and scrubby native trees looking for a Long-eared Owl, while Denver Holt keeps an eye out for the bird from the banks of the draw. It's hard, gnarly work, and Hernandez warns me to steer clear of the burdock, which has big round burs that will stick in your hair, snarling it hopelessly. On the ground, we find fresh whitewash, and nearby, the fresh remains of a pocket gopher, along with three or four pellets packed with the skulls of voles. An owl was clearly roosting here recently, but now it has vanished.

This scrappy, woody draw seems unlikely habitat, but the Long-eared Owl is partial to just this kind of gully, dense with thickets of hawthorn, Chokecherry, Russian Olive—places inhospitable to humans and to most predators, which is the point. The owl knows this as a singularly good, safe spot to roost through the winter, and it does so in a very unusual way.

Holt has been studying these owls and their roosts for thirty-five years. He has found roosting sites in five major draws here and has banded some 2,000 Long-eared Owls. This particular draw is special. For one thing, it's surrounded by good owl habitat. On the way there, we pass through rodent-rich fields of bluebells, rabbitbrush, bunchgrass, Bitterroot, and Prairie Smoke. Meadowlarks sing from the fence posts. Above us, a Golden Eagle soars, and we hear the alarm calls of ground squirrels. In the winter, this path is known to Holt's team as the "Big Walk" because it's brutal in the snow. But it's worth the trek. "No one thinks much of this habitat," says Holt, "but this is where everything hides out." He points out a low cluster of branches beneath an umbrella of dense hawthorn thicket. The bark is worn smooth from the talons of many, many owls that have chosen this place to weather the long Montana winters.

During the day most owls hide away in roosts to rest in safety after a long night hunting. Screech owls creep into dark tree cavities. Great Horned Owls seek perches high up in densely needled conifers. Some owls take shelter in caves, like barn owls, Little Owls, and masked owls. The Spectacled Owl rests in the canopies of rainforests and gallery woods, where predators are few. Male Powerful Owls roost in high trees, sometimes atop their dead prey. Saw-whets will nestle in almost any kind of refuge, as long as it offers impenetrable protection—snarls of honeysuckle and dense groves of rhododendron or thickety pitch pine.

There, sheltered from predators and mobbing birds, the owls will sleep if they can—sometimes with one eye open, says Niels Rattenborg, a neuroscientist at the Max Planck Institute for Ornithology who studies how birds sleep. Owls may be high up on the food chain, but they, too, have things to worry about. Rattenborg once observed a sleeping Eurasian Eagle Owl on a webcam. "Even though this is the biggest owl in Europe, the owl often slept with one eye open, and kept that eye facing away from the cliff wall, as if watching for threats," he says. It's not clear whether owls sleeping with an eye open are experiencing sleep in only half the brain the way some birds do. During this so-called unihemispheric sleep, one cerebral hemisphere is awake while the other slumbers, and the eye connected with the awake hemisphere stays open to monitor the world, allowing a bird to stay alert to danger and adjust to changing environmental conditions. Some birds, such as Great Frigatebirds, even nap in this one-sided way while flying.

Rattenborg says that unihemispheric sleep has never been observed in owls, which might have something to do with the way their front-facing eyes are wired to their brains. In other birds, each half of the brain receives input from the opposite eye. In owls, each hemisphere

Northern Saw-whet Owl sleeping

receives visual input from both eyes. "So, when owls sleep with one eye open, both the left and right sides of the brain receive similar amounts of visual input," he says. "In theory, this may cause both hemispheres to sleep less deeply than when both eyes are closed," but this has never been proven.

Holt says that his observation of wild owls on live cams makes him wonder if they sleep much at all. "Females appear to be awakened at any noise and very often." But owls do need sleep, perhaps as much as we do, and will sleep for long periods when they can. If they're deprived of sleep, they experience lapses in attention and cognitive acuity just as we humans do and suffer similar negative physiological effects on their immune systems, metabolism, and regulation of hormones.

Most owls roost alone or in pairs. But during the fall and winter, Long-eared Owls buck tradition. They actually huddle up. In that favorite sheltered spot Holt pointed out, eight, nine, ten Long-eared Owls will tuck in right next to one another. I think of owls as solitary creatures, jealous of their territory, hiding away by themselves by day and hunting alone at night. But here's a species that can be decidedly gregarious.

Not long ago, Holt and his team installed a solar-powered camera trained on this popular roosting site to get a view of the secret lives of these group-roosting owls. It was a first, and it's opening a new window on owl roosting behavior. How do the owls position themselves? Are there individual preferences for roost sites? Is there a hierarchy, with dominant birds getting the best, most protected roosting perches? The camera gives the team a chance to see the relaxed posture of roosting owls and how their behavior changes if a threat is present.

Recent film footage caught as many as thirteen birds in this one roost. Why would owls choose to roost together in such tight quarters?

For one thing, there's the frigid cold, says Holt. "Long-eared Owls cluster or roost in close proximity for thermal reasons." In general, larger owls are more resilient to cold than smaller owls. Great Horned and Great Gray Owls are large and well insulated with downy feathers. Big-bodied Snowy Owls are especially well adapted to frosty temperatures, with feathered feet and a dense layer of insulating feathers so effective they're second only to Adélie Penguins of Antarctica at retaining body heat, allowing the owls to endure temperatures down to −40 degrees Fahrenheit. These insulating feathers make Snowies North America's heaviest owl, about four pounds, twice the weight of a Great Gray Owl. But smaller birds like Long-eared Owls aren't built to withstand the bitter cold, so if they choose to stay in a habitat through the winter, it makes sense for them to huddle.

There may also be safety in numbers. This is the "vigilance" or "many-eye" theory. The more eyes on the lookout for predators, the better.

And there is a third, perhaps more intriguing reason. Shared roosts may make it easier for Long-eared Owls to find partners come time for the mating season. When Holt began this study, scientists thought these communal roosts were composed of family groups. "Turns out they're not," he says. "The birds are completely unrelated." The winter roosts may be a spot for the birds to find mates and then stay to breed locally for the next breeding season.

Year after year, the owls use exactly the same roost sites, even the very same perches. It seems reasonable to assume that the same birds were returning to use the same roosts. But when Holt and his team banded the owls, they discovered that it was almost never the same owls that roosted at the site from one year to the next. "Winter roosts in the exact same location were being used by completely different owls," he says. "These are just the best roosting spots, the best habitat with the best structures, and the owls know it."

You need only look at some of the current roosting sites of Sooty Owls on the rocky ledges and cliff overhangs of East Gippsland in Australia to see how faithful owls can be to a good roosting site, generation after generation. Large deposits of fossilized small mammal bones have been found at these sites, the prey remains of Sooty Owls that have roosted there for hundreds, if not thousands, of years.

The number of Long-eared Owls roosting together in Montana tops out at around twenty. Halfway around the world is a place where these owls gather in mind-boggling numbers.

Long-eared Owls roosting together in Serbia

If you want to see a spectacle of roosting Long-eared Owls—a true parliament—the best place in the world is the central square in Kikinda, a small town in northern Serbia near the Romanian border. It seems an unlikely spot for the planet's largest gathering of any owl species. But each year from November to March, hundreds of Long-eared Owls roost each day in the trees at the center of town. Wander into the main square of Kikinda, with its Austro-Hungarian buildings painted salmon and yellow, turn your gaze up into the pine, juniper, spruce, lime, birch, and poplar trees lining the square by the Orthodox church, and there will be hundreds of pairs of these lovely black-and-orange eyes looking down at you, says Milan Ružić. Look up into a single tree, and you'll probably see more than twenty or thirty owls.

The Serbian ornithologist has been studying these birds and their massive roosts since 2006. That year, Ružić and a small group of volunteers began a campaign to survey more than 400 villages and towns in northern Serbia in winter and were astounded to find large numbers of Long-eared Owls roosting in urban areas from Belgrade all the way to the Romanian border. They mapped as many of the roosts as they could find and counted the numbers of individual birds. "In sixty days of fieldwork, we counted twenty-four thousand Long-eared Owls in hundreds of different roosts," says Ružić. Over the next years, that number rose, and eventually, he and his team estimated that close to 30,000 birds were roosting in the region's villages and towns. Ružić's record for most roosting Long-eared Owls in a single tree during this fieldwork: 145. And at one location on a single day: 743. "That was a world record then," he says. "It still is today."

Urban naturalist David Lindo has led dozens of bird tours to Kikinda. "I'm now considered a 'son of Kikinda' and referred to as David Lindovich," he says with a smile. "But I remember going to Kikinda for the first time and my jaw just dragging on the floor. Looking up in that square and seeing so many owls—it was like walking into a Harry Potter set. In four days I think we saw upward of five hundred owls. To see all those owls en masse like that, it just blew my mind."

"No matter how experienced you are with nature and birds," says Ružić, "the first contact with a big group of owls in a roost takes your breath away. You think, 'Oh, this is wrong.' It's like a tiger walking down the middle of the road—you think, 'All these owls don't belong here; they should be out in the woods.' From when we're very young, we're taught that owls are just 'way out there.' You might be able to hear them, but you can't see them. And then you see an urban roost like this, and you realize that all the stories, all the things you've been told, are wrong. You get so excited. It's like a good natural drug for humans—it keeps you really high."

Long-eared Owls roost communally in other European countries, but not in the numbers they do in Serbia. In Germany and the Netherlands, the roosts hold around twenty-five birds, and in Hungary and Romania, as many as two hundred. There used to be big roosts in Slovakia and elsewhere, but when all the money came in to develop agriculture, the numbers of owls dropped. There used to be sizable roosts in the UK, too. One winter, David Lindo saw a roost of seventeen birds in West London, "but that was more than thirty years ago," he says. "We rarely get more than four together these days."

Why the birds are here in Serbia in such great numbers has to do with agriculture, specifically with food supply for the owls and the agricultural methods that sustain it. The whole region of northern Serbia is agricultural, and farmers avoid rodenticides and use more old-fashioned methods of harvesting that leave plenty of grain on the ground. In a typical alfalfa field, you can find millions of rodents, which makes the area excellent hunting grounds for the owls, says Ružić. He has calculated that Long-eared Owls throughout the region eat about thirty million rodents in a period of a few months. That's good for the farmers, he says. But for a long time, they didn't see it that way.

When David Lindo first observed the owls of Kikinda, he asked Ružić why more people don't know about them. Ružić replied that he and his team had been spending their time researching the owls, their habitat requirements, what they feed on—"and also convincing the people that live there not to kill them."

Back in 2007, Ružić recalls, "a lot of Serbians believed superstitious things about the owls. 'If there's a Little Owl on the top of the house calling, somebody in the house will die,' that sort of thing." He found instances of people harassing the owls, cutting down trees to try to get rid of them, even shooting them. So he began a massive campaign to educate people through conversation and media, giving them facts and figures. He engaged villagers:

"Okay, so these owls are calling during the breeding season for two and a half months in your village. How many people died during that time?"

"Well, no one."

"Okay. And for how long do you think the owls have been around?"

"Well, for at least twenty years."

"So how come anyone is still alive? We should all be dead."

In two or three years' time, Ružić could see positive change in communities around the region. Today, if you try to walk in Kikinda, you're stopped by dozens of people who want to show you the best tree in town with the biggest number of owls," he says. "The local community has become really proud of what they have."

The main square of Kikinda has been declared a nature reserve to protect the owls' habitat, the first ever urban protected area in Serbia. During Christmas and New Year's, no decorations are allowed in the trees around the square—which is a big deal, says Ružić, because those are the best places to display them. Anyone caught disturbing the owls during the festivities at Christmas can be fined up to €10,000. And the whole month of November is dedicated to owls. It's called "Sovember," after *sova*, the Serbian word for owl. Schoolchildren ages six to nineteen get involved writing poems, creating artwork, and making cookies. Kindergartners come in from villages around the town to see the owls and participate in educational programs dissecting pellets.

"To see the attitude change like that over the past ten years—it's one of the things I will take with me when I leave this planet," says David Lindo.

Ružić is still working to educate people and make them more aware of the owls. There are still villages in the region where people walk around beneath the birds and never look up, he says. Not long ago, he took a group of children from a kindergarten out to count owls in a village. "A lady came over to ask what we were doing. 'We're counting

Cluster of Long-eared Owls in Serbia

owls,' I told her. And she said, 'I've never seen a live owl.' And I said, 'Okay, at this moment, there are three hundred twenty-five owls around your head.'"

Owls roosting together in such numbers and in such proximity to humans offers some extraordinary opportunities for observation. "There's no better place in the world to come and see an owl up close than here in Serbia," says Ružić. "Sometimes you get an owl sitting up in a tree almost within arm's reach. They're completely used to humans. And if you just stay around long enough in the daytime, you can actually see their faces change."

In his talks, Ružić shows a panel of twelve photographs of the heads and faces of Long-eared Owls from Kikinda. The owls are so wildly different in their appearance that audience members ask whether they're different species, or at least subspecies. Ružić answers that they're just two or three different individuals in the same tree from the square. The photographer was sitting below them, pointing his camera up, catching them in different owl "moods."

"When a Long-eared Owl is relaxed, its face is rounded, its eyes are closed, and its tufts are down," he says. "That means, 'I'm kind of sleepy, I'm resting.' If they hear a sound or something, their face goes a bit more upward and they lift their ear tufts a bit. But if they get really alerted, they go slim and look twice as tall as usual."

"That's the great thing about this roost," says Lindo. "It's all the same species, but there are so many different looks to these owls. Some are puffed up. Some are thin looking and very alert with their ears erect. One or two look like David Bowie with one eye yellowish and one eye orange. They've all got character, and I just never tire of seeing them, never."

It's also possible to witness some fascinating behaviors. If a cat walks by or a human does something that disturbs some of the owls, they will stop their normal chittering to one another, says Ružić, "and they'll do this really funny alarm call. Then all of the owls start doing the same call. It's like, 'What's going on here? What's the trouble? Where's it coming from?'"

Individual birds sit on the same tree, the same branch, the same little stretch of branch, for more than ninety days. "They have their really tiny, really favorite spot," says Ružić. "But do they all come back to the same spot? We don't know." And do they have a hierarchy of best spots? He suspects they might. The owls change places within the roosts when the deciduous trees lose their leaves. Around 90 percent of the owls roost in birch trees, he says. But as fall progresses and the leaves drop, they slowly shift to conifers, and Ružić thinks it's the most experienced ones that do this first and get the best spots. Among Rooks in Europe, which nest and roost in colonies, the oldest and most experienced birds go to the very top of the tree, he says, "and they poop on the other birds below to show them 'this is my spot.' So this could be the case with the owls, too, but we don't know."

Nor do we know why the Long-eared Owl is the only owl species to

Burrowing Owl with black widow spider. Owls eat everything from insects to possums, rabbits, and young deer; some species can even tolerate venomous prey.

Snowy Owls bringing in lemmings to feed their young. By the time a Snowy Owl chick is two weeks old, it can swallow a lemming whole.

Female Great Gray Owl feeding its chick bits of vole or other small rodent

Great Gray Owl chick climbing a tree. A branching owlet may sometimes fall to the ground and then climb back up the tree with its powerful feet.

Boreal Owl fledglings. Chicks of this species leave the nest about a month after hatching.

Powerful Owl fledgling, which may stay with its parents for many months and continue to be fed by them

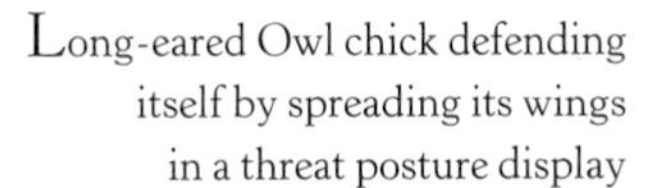

Long-eared Owl chick defending itself by spreading its wings in a threat posture display

Female Ural Owl swooping to defend her nest
in her Norwegian forest habitat

Great Horned Owl chicks nesting
in the protection of a tree crotch

Tropical Screech Owls (gray and red morphs) perched in a tree on the campus of the University of São Paulo

Short-eared Owl family in grassland nesting habitat

Long-eared Owl roosting under cover of a juniper tree

Tawny Owl landing on a snag nest in Norway

Images of owls abound in artwork from around the globe and through the ages.

Gold and agate brooch made in Italy, circa 1860

Egyptian relief plaque with face of an owl hieroglyph, 400 to 30 BC

A Greek owl "skyphos," a terracotta drinking cup from the fifth century BC

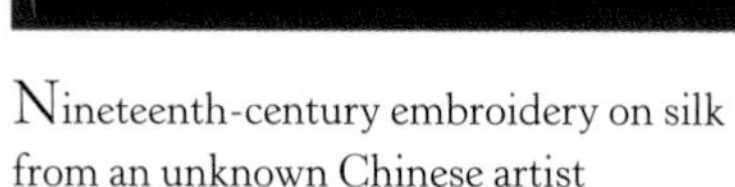

Nineteenth-century embroidery on silk from an unknown Chinese artist

Gold owl cap for a "staff" of a ranking official of the pre-Columbian Zenú culture, circa 1 to 1000 CE

Dependent Arising: Owl & Lemming, a sculpture in bronze by artist Terresa White. The sculpture stands in downtown Lake Oswego, Oregon, and, as White says, "celebrates the Yup'ik understanding of the interrelationship and spirit of all beings."

Spectacled Owl

Flammulated Owl

Mexican Spotted Owl

Eurasian Eagle Owl

roost in such numbers. Most owls won't tolerate proximity to nonfamily members. Denver Holt says Short-eared Owls sometimes roost communally, but individuals are usually spaced farther apart. The same is true for Snowy Owls, especially among the young once they've fledged, though again, they don't roost right next to one another.

From time to time, another species will tuck in with the Long-eared Owls. Ružić has seen Short-eared Owls among them and, on three separate occasions, a barn owl. "Short-eared Owls are closely related to Long-eared, so that's perhaps not surprising," he says. "But barn owls, wow!"

Why are these owls coming to roost in such numbers in villages and towns? Why aren't they roosting in natural areas? For one thing, like any good roosting site, villages offer shelter from weather and from predators. The winters are cold, and the winds are strong. The region is completely flat, and there are few trees, few forest roosts, especially after the broadleaf trees lose their leaves, so there aren't many places for the owls to hide or find cover. Fortunately, the villages and towns have planted conifers like spruce, black pine, Scotch Pine, juniper, and cedar for decoration in and around their streets and squares. After the owls feast in the farm fields, they seek shelter in the wooded squares and parks. In Kikinda, for instance, "the owls are really smart about positioning themselves in the trees right up against the local primary school," says Ružić, which keeps them warm and helps preserve energy. Also, the goshawks and eagle owls that will hunt them down in other settings won't come into the towns and villages.

As Denver Holt and others have suggested, the big roosts may be a place to find mates. "It's clearly an owl party where the owls get to meet many other individuals," says Ružić. "Especially if you're a juvenile or a young adult, it's much easier to find a mate." The roosts may also be information-sharing centers, where the owls share or glean information about food sources. The theory that birds in roosts or colonies might gain information about where to feed is known as the information center hypothesis. Scientists have documented it in only a few species, including Common Ravens, Ocellated Antbirds, and Cliff Swallows. But both Holt and Ružić think there might be something similar going on with communally roosting Long-eared Owls. "The owls may be passing along information about good hunting spots," says Holt, "whether it's sharing information or stealing it."

Ružić has observed owls in the roosts setting off for a hunt and believes their approach is highly targeted. "In the evening, they fly out of the roost to a nearby treetop, and it's like they're thinking, 'Okay, let's look around and think about where I had that really good juicy mouse last night. There's food all around the place, but I need to be quick. I need to be efficient. I should go east.' Sometimes I think they have a mental map of the region with the good places to hunt," he says. "Then you see one owl going off in that direction, then another follows, then five more take off, going in the same direction." That looks like information sharing to him.

Ružić also suspects that the roosts themselves may be signaling information to migrating owls passing over that this is a good staging ground, a stepping stone on migratory pathways from Scandinavia to southeastern Europe. "Think of being a young Long-eared Owl. You've been pushed away by the cold blowing winds from Siberia or northern Europe. You're coming south in the night, and you're in a completely unknown area. You know nothing about the habitat. You know nothing about food sources. And suddenly you see loads of other Long-eared Owls roosting, fluttering around, foraging, and you say, 'Okay, there must be food and safety there.' So you join them for a month or two. That may explain why our numbers are so big," he says. "All the activity around the roosts is attracting Long-eared Owls from all over Europe."

Ružić sees the roosts as learning centers in another way as well. To roost in an urban environment, a young owl must learn to overcome its natural fear of humans. It shows up in the middle of town with the sky lit up by artificial lights, says Ružić, and there's noise and traffic and the buzz of people walking all around. "Visit a roost in October or November, and you can easily tell which birds are older and more experienced because they pay almost no attention to humans," he says. The newcomers, on the other hand, "are really nervous for the first couple of days, trying to pick out whether any sound, any movement is a threat. 'Should I leave or not?' So they learn from the owls who have been around—'Yeah, that's just a guy walking below or riding a bike. That's just a dog barking down there. No need to fly—it's safe.' This is how they learn from each other what's dangerous and what's not," he says. "It's always about learning."

OR TO GO . . .

When it comes to owls, there's probably nothing more astounding than seeing hundreds roosting collectively in a single place. But if there is, it's the thought of multitudes of owls flying silently above our heads through the night sky.

It was once believed that owls were mostly sedentary, remaining more or less in the same place all their lives. That's true for some species—the Little Owl, Barred Owl, and Northern Spotted Owl. But other species migrate: they move around for the same reasons other migratory birds do, to find food. But when it comes to the details, owls are hard to pigeonhole. They seem to defy tidy categories: This species is a migrant. That one is nomadic. This one is "irruptive," meaning the bird periodically migrates in unusually large numbers from its breeding grounds, sometimes—but not always—because of a scarcity of prey. With owls, it's more subtle and nuanced.

"These birds have been around for millions of years," says David Johnson, "so we might expect them to have evolved many ways for dealing with food limitations." Their strategies vary wildly between sexes, among age groups, and under different environmental conditions. What applies to one species does not necessarily apply to another. Some species, such as the Long-eared Owl, Short-eared Owl, and Burrowing Owl, are seasonally migratory. So is the Eurasian Scops Owl, which makes long migrations from Europe to wintering grounds south of the Sahara. Even within a single species, there can be tremendous variation. Snowy Owls, for instance, use every migratory strategy in the book.

"We're far behind in our understanding of the migratory patterns of owls versus other birds," says Scott Weidensaul, an expert on bird migration, in part because they're nocturnal and just so much harder to study.

Thanks to the efforts of a flock of dedicated researchers, two species—one tiny and elusive, the other huge and showy—are starting to yield some of the secrets of their surreptitious journeys and, in the process, illuminating how owls move and why.

Halloween. It's a cool autumn night with clear skies and gentle winds from the north-northwest. No moon to speak of. Good conditions for a migratory push of Northern Saw-whet Owls. I'm headed to a banding station in Powhatan, Virginia, to witness the capture and banding of these alluring little owls.

Some people know the Northern Saw-whet from its limelight appearance in the seventy-five-foot Norway Spruce from upstate New York that was to serve as a Christmas tree in Rockefeller Plaza in 2020. It was during the pandemic, and the bird became a symbol of resilience, stealing the hearts of New Yorkers and the whole nation. "Rocky" was

Northern Saw-whet Owl roosting

found tucked in at the base of the tree. She undoubtedly got trapped there when the tree was bagged prior to cutting and transporting. Rocky was cared for by a wildlife center before she was released back into the wild. How she made it all the way from Oneonta to the city in that downed tree remains a mystery—like so many other things about this species.

The highly nocturnal and secretive Northern Saw-whet occurs in forests across North America. Small as a robin, with a heart-shaped face, oversize head on a squat body, and bright, fervent eyes, the saw-whet is arguably the world's most adorable owl, bursting with moxie. It looks more like what you would think a baby owl should look like than most real baby owls do. But don't let its tiny size and baby face fool you. That little beak and those needlelike talons can easily rip off the head of a deer mouse or vole.

Jim Duncan tells a story that reveals the nature of this little bird. He and his wife, Patsy, were surveying for owls in southeast Manitoba one spring. They heard a saw-whet whistling, "a male just singing his heart out," says Duncan. "And I said to Patsy, 'Gee, I wonder . . . if we imitate a Great Horned Owl call, will the saw-whet be intimidated and stop calling?'" He hooted like the big owl, and the saw-whet instantly went quiet. "I turned around to Patsy and said, 'Wow, I guess those owl survey recommendations about not playing big owl calls first were right.' Two seconds later, the saw-whet hit my head, took my wool hat right off. It's a little owl with a really big attitude."

Because of their nocturnal, reclusive nature, saw-whets were once thought extremely rare. Wherever they did reside, they were thought to stay put. For centuries, the abundance of saw-whets migrating in the autumn simply went unnoticed. But in the early twentieth century, keen-eyed observers at Lake Huron and Lake Erie changed that. One fisherman on a steamer crossing Lake Huron in October 1903 reported seeing a large migration of small owls fitting the saw-whet's description, many of which landed on the boat. A few years later, a note in the ornithological journal *The Auk* described a natural disaster on Lake Huron's shores, when large flocks of migrating birds crossing the lake were overwhelmed by snow and cold, fell into the water, and drowned. Close to 2,000 dead birds washed up on a two-mile stretch of shore—Dark-eyed Juncos, Song Sparrows, thrushes, kinglets, wrens, and twenty-four Northern Saw-whet Owls. "The Saw-whets were a surprise," wrote the author. "They are rare in western Ontario, and one sees them only at intervals of many years. Evidently they migrate in considerable numbers."

The matter was further illuminated in the fall of 1910 when Canadian ornithologist Percy Taverner and a birder friend found a couple of dead saw-whets on Point Pelee, a long peninsula of marsh and woodlands that tapers into Lake Erie. They decided to search for live ones in a small thicket of red cedars. They spotted one owl, then another with a mouse in its bill, which "rose high up on its legs, leaned forward, and glared at the intruder, still holding the body of the mouse in its bill, exhibiting no fear and . . . following every movement with its golden eyes." Though the saw-whets were hard to detect, "all quiet and so near the color and contour of other natural forms as to be most inconspicuous," the pair discovered a dozen of the little owls. If a small thicket yielded so many saw-whets in just two hours, the total number of this rare species on the Point must have been very great, they wrote. Then came the clincher. The next day, they couldn't find a single owl. "We worked the whole end of the Point with great care, but except for the scattered remains of another unfortunate, saw not a sign of them. They had evidently departed in the night."

Now we know that Northern Saw-whet Owls move. At least some of them do. At least in some years, some of them do. And every four years or so, a lot of them do. But not always. What dictates their irregular movement patterns, where they go, when, and why is a complex riddle.

Sorting all of this out is the work of Project Owlnet, a collaborative research project that is the brainchild of David Brinker, an ornithologist at the Maryland Department of Natural Resources. The project started in the 1990s to coordinate a small group of banding stations in eastern North America. The plan was to coax as many owls as possible from the night sky and to band each with a unique identifying number so that if it was recaptured, researchers could track and record its movements, with the ultimate goal of understanding the timing and pace of the owls' movements and identifying important habitats and routes.

Now Project Owlnet is a network of more than 125 banding stations across the United States and Canada run by dedicated crews of hundreds of volunteers. Every fall they tend mist nets and band thousands of saw-whets all through the night to try to solve some of the species' many mysteries. (The same night I was out banding in Powhatan, a team of volunteers at the Owl Research Institute, some in ghoulish costumes, were celebrating their tenth year banding saw-whets for the project.)

The approach taken by these banders and others around the country minimizes the impact on the birds. "While the banding of any owl is a disruption to their night," says longtime bander Dave Oleyar, "it results in few to no long-term impacts and is an important tool to help shine a light on movements and also on survival rates—the big black box in wildlife studies. Many of the birds we release after banding sit in your hand for a minute or two to get their bearings, fly up to a nearby tree and check their new jewelry, and then go back to their business."

The Powhatan banding station is managed by Julie Kacmarcik and Kim Cook. That Halloween night, Cook and another volunteer, Diane Girgente, are running the show, the three of us masked up for COVID. The station is a small cinder block building in the woods equipped with tools for measuring and weighing the owls and abundant owl decorations courtesy of Kacmarcik, including a BEWARE OF ATTACK OWL sign beneath the clock.

Around 8:30 p.m., we're on our first run of the night to check the mist nets a mile and a half up the road. There's eyeshine on the road ahead, an opossum crossing. The banders conduct hourly net checks to make sure that if an owl is caught, it's extracted from the net right away so that predators such as opossums, raccoons, foxes, or big owls don't reach it first. This back-and-forth routine is repeated all night long.

As we walk in the dark down a rough path through mixed hardwood forest, our headlamps catch little winking green lights on the ground, thousands of them—the eyes of wolf spiders. They remind me, along with that opossum, that this night world is *their* world, the kingdom of insects, spiders, small mammals, and birds that do their deeds in the dark.

When we get close to the nets, we can hear the Foxpro emitting the shrill, repetitive *toot, toot* advertising call of the male saw-whet. It's not the kind of call where you think, "Now *that* is an owl." It sounds more like the warning beep of a truck backing up, only faster. In the 1980s, owl banding stations didn't use audio lures. Instead, they put up mist nets at night and hoped to catch the owls en route with this passive method. But in 1986, an enterprising ornithologist in Wisconsin began using a tape recording of the male saw-whet's advertising call and managed to capture ten times the number caught with passive netting. Clearly, the call was attracting owls from considerable distances. Now it's regular practice to use the audio lure, and today, more Northern Saw-whet Owls are banded in North America than any other owl species.

No luck this time, however. The mist nets crisscrossed in an X in the dense understory are empty. This is the way it goes with saw-whets. Good runs and disappointing runs, years when the owls are few and far between, and years when they literally flood the nets and keeping up with the steady flow of birds is almost impossible.

Julie Kacmarcik has run the station since 2006. That first year, she went for a whole season shuttling back and forth between nets and the station every hour, night after night, from November to March, with not a single owl to show for it. In between net runs, she stayed up knitting a never-ending scarf. During the day, she worked as a clinical coordinator in the emergency department at a major teaching hospital. At the ED, her days were full of trauma and illness. The owls, she thought, would be a relief.

"That first year we caught zero owls, and I remember our master bander Bob Reilly saying, 'Well, Julie, no data is still data.' And I'm like, 'Well, Bob, that's not the data I want.'"

The following year, Kim Cook became Julie's partner in crime, and on the very first run of the very first night, they caught two owls. "Talk about the happy dance. We were just ecstatic," Kacmarcik recalls. She was hooked, and she found that the owls were indeed an exhale from her day job. At the ED, she and her team gave everything they had to try to resuscitate a patient—and often failed. "So many times, that person—a brother, a mother, a friend—didn't plan on coming in to see us that day," she says, "didn't expect their life to be ending." It was wrenching. For her team's sake, she brought the concept of "The Pause" to the controlled chaos of the ED, a moment of silence after the death of a patient, a time-out for everyone involved to be quiet and reflect, to process.

"I definitely have compassion fatigue," she says. "It's hard not to take it home with you. The owl work is restorative, life-giving, the exact opposite of my job," she says. "There's nothing like a night out with the saw-whets to erase a horrible day in the trauma room. I don't want to say it takes it all away. But sometimes I'll say to Kim, who I've banded with for a long time now, 'You know, I could really use an owl right now.' And she gets it."

On a typical fall night, Kacmarcik is there to check the nets every hour. In a good year, she says, she has seen many a sunrise. The year 2012 was a very good year. Kacmarcik and her team caught 169 owls, "which for some stations, like up in Ontario, northern New England, is nothing," she says. "But for us, it was incredible. In one net run, there were nine owls. I was at one net taking out an owl, and another one blew by my face and hit the net, which was crazy." There were stretches that year when Cook and Kacmarcik stayed up three or four nights in a row. "We'd throw down a cot and take catnaps. When it's really good, you just don't want to quit. Many nights we left the field station, went home and took a shower, and went off to work."

At around 9:30 p.m., on our second net run, luck is with us, and we see from a distance a small brown body upside down and wriggling in the net. Girgente flips off the Foxpro, and the night goes quiet. Cook reaches into the net and grasps the owl's small, feathered feet so its sharp talons won't puncture her fingers, then finds its belly, untangles it, and slips it free of the mesh. It's a female. We can see immediately that the pupil in her right eye is misshapen, large, and more oval than round, possibly a birth defect or the result of an injury. But judging from her feather coloring, Cook says, she's an older bird and has done all right. Cook puts her in a small muslin bag, and I carry her back through the strange starry field of spider eyes.

At the field station, we take her out of the bag. Her feathers are velvety soft, her eyes a bright burning yellow. She's calm, weirdly so. Saw-whets are usually very docile, Cook tells me. "They seem almost mesmerized in your hand—at least the females are. Screech owls are feistier and will use their feces defensively. The smell is so strong you can't get it off your hands. Saw-whets, on the other hand, are clean, polite, nice to handle."

The banders don't wear gloves. If they did, they might risk injuring the bird. With gloves on, you just don't have a feel for how much pressure you're exerting, you don't have a sensation of the strength of your grip, Kacmarcik told me. "The important thing is just knowing where those legs and talons are. You may get some nipping and biting, but the part that's going to get you if you don't hold it correctly is going to be those needlelike talons."

Cook slips a tiny metal band with an identifying number onto the bird's leg and crimps it, making sure it's not too tight or caught on any feathers. She and Girgente stretch out a wing to show me the ragged trailing edge of feathers that help to quiet the little bird's flight. They measure her tail and her wing chord—its length—which helps determine her sex. Birds are sexed based on a combination of wing chord and mass. Females are substantially larger than males, weighing on average 100 grams (about the same as a bagel—without the cream cheese, says Kacmarcik) compared with a male average of 75 grams. But the truth is, there's little doubt this bird is a female.

One of the telling aspects of the data turned up by Project Owlnet is the dramatic overrepresentation of females caught at banding stations across the country. They outnumber males four to one. In one study of more than 40,000 saw-whets caught at 252 banding stations over an eight-year period, 86 percent of the owls were female and only 14 percent were male. Why? Certainly the male courtship call used in the audio lure is more attractive to females than males (even in the nonbreeding season). That "advertising" toot suggests to females that there may be a mouse to be had from an obliging male. And it suggests to males that the territory is taken, driving them away. But that's not the whole story.

This female is completely still in my hands, almost limp, and squinting. When she opens her eyes wide again, we compare their color with a little Benjamin Moore paint chart, the official banding scale for eye color: Oxford Gold, Bold Yellow, Golden Orchards—her color. We measure her bill and check around her keel, or breastbone, for muscle mass, and under her wings for fat stores. Sometimes you get a butterball, says Kacmarcik. "Our heaviest was a bird we caught one March, so she was on her way back to her breeding grounds. She had huge fat stores and weighed 122.6 grams—off the charts. And I remember she looked so roly-poly, like she had no neck."

Don't all owls look like they have no neck?

"This one was actually *round*," says Kacmarcik. "She was totally like a ball. And it was neat to look in her wingpits and in her furcular hollow and see yellow fat bulging. I thought, 'Yep, that's good. You're good to go up north to breed.'"

Our bird has only little strands of fat in her wingpit, what you would expect during fall migration, when she's burning off what fat remains after the breeding season. To age her, Cook and Girgente look at her wings under a black light. The wing plumage of owls contains deposits of UV-sensitive pigments called "porphyrins" that fluoresce a bright pink when exposed to a black light. Porphyrins degrade over time, particularly with exposure to ultraviolet light, so the older the feather, the less brightly it fluoresces. For most owls, shining a UV black light on the underwing is a simple and effective way to assign age to an owl in its first few years of life. (Not in Burrowing Owls, however; they spend too much time in the bright sun, so their porphyrins quickly fade.)

"A young bird's feathers will appear uniformly pink," Cook explains, "whereas an older bird's feathers will have some combination of pale pink, white, and a dingy cream color, representing two, three, or more generations of feathers." Our bird is a complex mix of all three colors, so she's classified as an ASY, after second year.

All of these meticulous measurements and the bird's band number are submitted to the Bird Banding Laboratory, which maintains a database of all birds banded in the United States. When a banded bird is recaptured by another bander, the lab sends the details of the recapture to the original bander. The records are then used by Project Owlnet in its continent-wide effort to monitor the owls' movements. "The most exciting part of the saw-whet banding is when you capture an owl that already has a band on it," says Kacmarcik. "Our longest-distance recapture was from Manitoba, which just about made me fall out of my chair: Manitoba to Powhatan, that's an incredible distance, about fourteen hundred miles."

It's because of Cook and Kacmarcik and the flock of banders across

Northern Saw-whet Owl in flight

the country that we know something not only about the migratory routes and mileage of these birds, but also about their numbers. We know that a big spike in population tends to occur roughly every four years. The boom is tied to the abundance of small mammals, especially red-backed voles, in the boreal forests of Canada, northern-tier US states (Wisconsin, Minnesota, Michigan), and New England, which are the breeding grounds for saw-whets that migrate down the East Coast. This in turn depends on the crop of cones produced by the conifers in these regions. A banner cone crop translates into a banner year for voles, which results in a boom in baby owls, and the young move south.

The big years for saw-whets boggle the mind and fill the nets. In 2012, a Project Owlnet team managed by Scott Weidensaul was running six experimental stations in Pennsylvania, in addition to the usual three banding stations, to try to figure out whether the owls were moving on a broad front across the landscape, the way most nocturnal migrants do, or whether they were following forested corridors along ridgetops, the way diurnal raptors do. (It turns out they use both strategies, depending on the weather.) Those six additional stations ran all night, every night, with the help of a dozen young research technicians. "And we caught like four thousand saw-whets that fall," says Weidensaul. "It was insane. At one net check at one station, there were twenty-seven owls, twenty-seven little bags hanging on the clothesline."

The owls in those nets were mostly female, and the males were mostly hatch-year birds. "In those big years it's a big slug of juvenile males coming in," says Weidensaul. "Over the past twenty-six years, we've banded more than twelve thousand saw-whets in Pennsylvania, and we've caught exactly a dozen adult males—literally one in a thousand."

Cook tucks our female owl back in the muslin bag and places her on the scale and then, while she's still in the bag, takes her into the dark back room for a half hour or so to acclimate to the dark. Finally, it's time for her "soft release." Cook sets the little bag on a fencepost with the owl upright inside and gently peels off the bag. The bird looks around, flies up to sit in a cedar tree for a few moments, and then goes off into the dark to continue her journey.

That night, I dream of releasing a saw-whet in broad daylight in a large park. The owl flies from tree to tree, edging the park, but she's always there, always visible.

The "always there" part could be considered accurate. The "always visible" part is very much a dream.

"The biggest thing we've learned from Project Owlnet is the utter and complete ubiquity of Northern Saw-whet Owls," Weidensaul told me. "I would argue that these owls are the most common small forest raptor in North America. When I started working with saw-whets in the mid-1990s, they were the symbol of the state Wild Resource Conservation Program and a candidate species for the state endangered species list because everybody thought they were really, really rare. And it turns out they're just really, really *rarely seen* because they're so highly secretive and highly nocturnal."

A case in point: For years Weidensaul and his team used radio telemetry to try to learn more about where saw-whets hang out in the day when they're migrating, with limited success. "We were interested in their stopover ecology, what kind of trees they were roosting in, etc., so we were catching them, putting on small radio transmitters, and then tracking them down in the daytime to see where they were roosting. It was easy to find the tree the saw-whet was roosting in," he says. "It was hard to find the saw-whet in the tree it was roosting in. There have been more times than I can count where I'd be going around and around in this thicket where the transmitter signal is coming from. And I know there's a saw-whet in there somewhere, and I'm peering in, and, finally, I'm like, 'Okay, it obviously either dropped the transmitter or something killed it and picked the transmitter off.' And then I step in and flush a saw-whet because it was sitting in there the whole time. It just won't move."

Moreover, says Weidensaul, the birds tend to roost in "supercanopy" hardwood trees, "in the outer branches, with their head stuck up in a clump of leaves. After hardwood trees drop their leaves, the owls move into evergreen cover, conifers or honeysuckle or rhododendrons or big blowdowns. "In any case, they're really easy to miss."

But here's the thing, he says. Pretty much anywhere in North America south of the subarctic tree line, if you set up an audio lure and a mist net and have a little bit of patience, you're going to catch saw-whet owls.

Given how ubiquitous they are, there's a lot we still don't know about these little owls. Simple things, like how high up in the air column they fly when they migrate. Weidensaul frequently gets asked by wind energy developers at what altitude the birds are migrating, because they're concerned about owl mortality at ridgetop wind turbine sites. "And the fact is, I can't tell them," he says. Transmitters with the capacity to measure elevation are too heavy for small owls, so there's currently no way of knowing how high they fly.

Researchers are still trying to sort out some of the other basics, such as where the birds roost when they're moving and where they feed, and also, the more complex questions, where adult males go in winter, and why the irruptions of these little owls have lately become more unpredictable.

"You can fool yourself into thinking you know more than you know," says Weidensaul. When he and his team were using radio telemetry to study where these birds were roosting, they found that many were resting close to open water, especially running water, and often in places that had a dense understory. "So I thought, 'Ah, they're obviously picking roost sites where there's going to be an abundance of rodents, because all that dense understory would presumably have a lot of small mammals.' But when we started tracking these birds at night, we could see the owls moving around the landscape. And it turns out they never hunted anywhere near their roost site. They always flew a distance to hunt. We had one owl that would roost on top of a mountain and fly three miles down into the valley to hunt along the edges of a dairy farm in this wooded stream valley, and then it would fly back up to the top of the mountain and roost at night. Until we started tracking, we had no idea these birds were doing that."

Also in the category of "thinking you know more than you know" is the pattern of big flight years, or irruptions. In past years, Weidensaul has been able to predict the saw-whet numbers moving on the East Coast in any particular autumn based on the crop of Balsam Fir cones in southeastern Canada and New England, which provide food for mice and voles. He used to be able to make a pretty good guess: "The cone crop this fall usually tells me what the flight is going to be like next fall," he says. But lately that has changed. Since 2012, the predicted boom years have not materialized. He thinks that's because climate change in the past fifteen or twenty years has affected the cone crop cycles, which affect the number of migrating saw-whets. "David Brinker, who has been banding longer than I have, has a poetic way of putting it," he says. "With every one of these big flights, it seems like the tide doesn't go quite as far up the beach as the last one. So it's just not quite what it was."

One thing we do know, thanks to the efforts of banders: the speed at which some saw-whets make their journeys. Not fast, says Weidensaul. "Owls are kind of all over the map when it comes to their migratory pace. You've got species like Flammulated Owls that are high-intensity, long-distance migrants. They blast out of the Northern Rockies flying around 180 miles a day, and in a very short time migrate to the mountains of southern and central Mexico." Why such a speedy pace? "Because they're insectivorous. If they stick around too long, the insects die, they run out of food, and they're dead," he says. "Then you look at saw-whets migrating out of the same area, and they take this kind of lackadaisical, 'Oh, I'm just going to kind of poke along here' approach to migration. They're like molasses going south. When a colleague did an analysis of banding records on saw-whets a few years ago, the average daily migration distance was fifteen miles."

Their slow pace may reflect a difference in the nature of their movement. Though many saw-whets are faithful to their routes, they don't appear to be migratory in the conventional sense, moving between the same wintering and breeding grounds every year. They're more nomadic, at least in some parts of their range, moving from place to place looking for a good vole supply to get them through the winter. A team of Canadian researchers studying the abundance of saw-whets over time determined that by moving around in this way and evaluating vole supply as they go, the owls avoided years of widespread reproductive failure. While they're on the go in the nonbreeding season, they appear to be comparing relative vole abundance among various sites, looking for vole hot spots to return to, or to remain in, for breeding.

Why banders catch so many female saw-whets and so few males during their movements was once a mystery. But clues from the banding data are solving the puzzle and casting light on a more general phenomenon in owl migration.

When biologists Sean Beckett and Glenn Proudfoot studied the sex ratios of banded saw-whets, they found a revealing pattern: Adult males were caught in greater proportion at more northern latitudes. This suggests that these males "may be less migratory and remain farther north than females," the scientists wrote, which explains why banding stations would be catching mainly females, and why, over the past quarter century, Weidensaul and his team in Pennsylvania have caught so few adult males.

This sort of "sex-specific migration strategy" is typical of other owl species, too—including that close relative to the saw-whet, the Boreal Owl (or Tengmalm's Owl, as it's called in Europe)—and may have to do with differing food requirements between males and females. In Finland, owl researcher Erkki Korpimäki found that female Boreals (the bigger sex) migrate south from colder regions to areas with more plentiful food, and the adult males remain up north, closer to spring breeding territories. Females can't afford the risk of diminishing food supply in winter, so they seek a warm, safe place to fuel up before they breed again. Males want to stay where they have quicker access to the best territories in the spring. So they move in the winter, but they move laterally, shifting around looking for places with a big rodent population. Male saw-whets may be doing something similar. They're taking a chance staying in a cold environment, says Weidensaul. "If they have a really harsh winter and they don't find a place where there are lots of mice, they're going to die. But if they survive, they've already picked out a territory and they're waiting there when the females come back in February, March, April, and they can start calling and find a mate right away." It's the same strategy used by those Burrowing Owls in Oregon. In fact, this is turning out to be true in many owl species, says David Johnson—females move farther south, and males stay north, closer to breeding sites. "It's a risk the males take so they can enjoy the benefits."

No one has documented any of this yet in saw-whets, but Weidensaul is excited about a new program to deploy hundreds of nanotags with identification numbers on the little owls, tiny temporary radio transmitters that will be used to track the birds with the Motus Wildlife Tracking System. The transmitters send out signals every two to thirteen seconds, which are picked up by Motus reception towers staged along migration routes. The towers are linked to the internet and download the nanotag information in real time to track how an individual bird migrates. "We'll be tagging breeding adult pairs of saw-whets and juveniles on the breeding grounds and adults and juveniles in migration," he says, "so we can finally figure out where these birds—especially those missing adult males—are going."

Despite all the efforts to understand saw-whets, the little owls still stump science on fundamental questions, "questions we can't answer because the owls are too tiny to carry satellite transmitters," says Weidensaul. "You can't put a forty-five-gram transmitter carrying an altimeter and capable of communicating with satellites on an eighty-gram owl."

But you *can* put one on a big, sturdy Snowy Owl, which weighs an average of around 2,000 grams. Denver Holt was the first to put one on Snowy Owls, six birds in Utqiaġvik, Alaska (then known as Barrow), in 1999. Two of the birds flew from across the Chukchi Sea to Russia,

Female Snowy Owl with a satellite transmitter

summered along the northern coast of Siberia, and then arrived back on Victoria Island in Canada. Other researchers have followed suit, including members of the International Snowy Owl Working Group and Project SNOWstorm, teams of collaborative researchers seeking to understand Snowy Owls using innovative science.

What scientists have learned by tracking the owls has confirmed earlier observations and sparked new ones. If Northern Saw-whet Owls seem unpredictable in their migratory patterns, Snowy Owls are perhaps the oddest of all.

There's no mistaking the majestic white owl: larger than a Great Horned Owl, much heavier than a Great Gray, and—males, anyway—a pure, brilliant white with black-rimmed tigerish yellow eyes. Females are bigger, weightier, and as we know, more densely marked. Snowies breed across the open Arctic tundra in Russia, Scandinavia, Canada, and a small piece of Alaska. The big birds are known for their hunting prowess, their nest defense, and their powerful flight. Also, for their jaw-dropping consumption of lemmings, some 1,600 by a single individual in a single year.

For decades, observers have noted the bird's dependence on lemmings for breeding. Russian researcher Irina Menyushina studied the Snowy Owls of Wrangel Island for twenty-seven years. In the 1990s, she noted that the owls' reproductive success depended on the numbers of lemmings present and was fully realized only during peak lemming years. But only lately have scientists begun to understand the nuanced nature of the relationship between bird and rodent and how it affects the owls' movements.

Over the past 30 years, Denver Holt has studied Snowy Owls in a 100-square-mile area around Utqiaġvik, more than 300 miles above the Arctic Circle. It's the only region in the United States where Snowy Owls breed and raise young, and it's the home of the Iñupiat people, who have lived side by side with the owls for more than 4,000 years. Over the course of his study, Holt has analyzed 14,000 Snowy Owl pellets containing the remains of more than 43,000 items of prey and has found a diversity of animals represented there, 33 species of mammals and 20 species of birds. But in the breeding season, the Snowy's diet is 99 percent lemmings.

"It all comes down to lemmings," says Holt. When the little rodents abound in a nesting area on the tundra, everything, including the owls, thrives. But in low lemming years, the owls must move on to a new site or fail to nest at all.

These birds routinely move great distances in the Arctic, actively searching out their prey. "Males, females, young of the year, all have different strategies," says Holt. "And different individuals have different strategies." Some Snowy Owls are faithful to their breeding and wintering sites. Others disperse to new ones. Many of the birds will stay on their breeding grounds through the winter while others explore from winter to winter, traveling from Alaska to Russia and back. One radio-tagged Snowy traveled 600 miles in a single day. "They seem to know where they're going and remember locations with astonishing precision," David Johnson told me. "Tagged Snowies in the US have been known to come back to the same telephone pole within a winter territory they occupied six years earlier."

In looking for good breeding sites, the owls will search for long periods and travel long distances. The distance traveled and the duration of prospecting movements are longer in years when the density of lemmings is lowest, which makes sense. A 2014 study by Jean-François Therrien using radio transmitters showed that Snowy Owls in the Canadian Arctic seeking a place to breed would search for months and travel distances up to 2,500 miles. From one year to the next, females would nest in sites on average 450 miles apart.

The Snowy Owls of Alaska and Canada sometimes head south to winter in the prairies of southern Canada or the northern-tier states of the United States. They've been known to fly much farther south in a big breeding year. Every so often, when lemmings abound in the Arctic tundra, there's a boom of young Snowy Owls, and birds five or six months old will move south in large numbers. "The long-standing myth is that these irruptions were driven by starving owls," says Holt. They're not. "Prey can be abundant, and the owls still move south." It's likely that other owl species, such as the Northern Hawk-Owl and the Short-eared Owl, behave in a similar way. This is the opposite pattern of irruption in species like Boreal Owls and Great Gray Owls, when adults move because of a depletion of their prey resources. "One of the only things we know with certainty," says Holt, "is that irruptions of Snowy Owls are the result of a boom breeding season somewhere in the Arctic, which results from an abundance of lemmings." He's quick to note that the phenomenon was first observed by naturalists and

Snowy Owl with a lemming

ornithologists of the 1930s and 1940s and later noted by trappers and banders.

What makes for a good lemming year? "It's hard to know," says Holt, who has studied the small mammals along with their big predators for decades. "There are so many influences—food resources, amount of snow, quality of snow. It's like the banner cone crop that occurs from time to time, when everything comes together with the right weather conditions, and you get a great crop. It's the same with lemmings. The populations fluctuate from season to season, but every once in a while, the conditions are perfect for a real surge in the numbers, and that results in high numbers of Snowy Owls nesting and producing young."

During an irruption, Snowy Owls can appear almost anywhere in big numbers, on the northeastern Atlantic coast or the northwestern Pacific coast, even in the Great Plains or farther south, says Holt. Ornithologists and naturalists have noted "heavy flights" or "invasions" since the 1800s. One of the most memorable recent irruptions took place in the winter of 2011/12, when owls were recorded in every Canadian province and in thirty-one states within the United States. During the irruption two years later, an abundance of young owls flew southward all the way to North Carolina, Florida, even Bermuda, captivating birders and Harry Potter fans alike. "But these are extreme records," says Holt.

Perhaps strangest of all, some Snowy Owls actually move *north* in the winter.

Why would a Snowy Owl do this, travel north into the perpetual depths of the Arctic winter darkness? "People living in and exploring the Arctic have witnessed this for centuries," says Holt, "and the Iñupiat and other native people knew about it before anyone." But scientists are now attempting to parse the mystery with satellite tracking and other studies.

When Jean-François Therrien was doing his PhD research on Snowy Owls in the Arctic in 2007 and 2008, he fitted the birds with transmitters to track their winter movements. In his data, he noticed something odd. The owls were not only passing the winters way up north, but also spending large amounts of time out over the sea. "These are terrestrial specialists," says Therrien. "They're supposed to be feeding on lemmings. And here they are, out on the sea ice two hundred kilometers from the coast. This is really surprising behavior for a small-mammal specialist!"

When Therrien presented the images to his supervisor, the response was, "Are you sure you know what you're doing? That doesn't look right. Go out and do it again."

Therrien did. But sure enough, the owls were out at sea. A closer look at the satellite images showed that they were hugging the edges of the ice near open water patches called "polynyas," which are created by currents and tidal movements and tend to persist or recur from year to year. "What were the owls doing there? Going for a swim?" Therrien asks jokingly. "No! But they weren't looking for lemmings either." The polynyas were packed with gulls, auks, and big sea ducks, including Eiders, which gather there in huge numbers in winter, fishing and feeding on mussels on the seafloor. "So what the owls are doing is sitting on an ice floe, drifting around, and preying opportunistically on these big waterbirds," says Therrien. An owl can prey on an Eider? "These are big birds, even heavier than a Snowy Owl. But yeah, they can."

Holt had shown this in his earlier pellet studies, in which he found the remains of pintails, shorebirds, seabirds, and the favorite wader in the Iñupiat region, the Red Phalarope.

This feasting on waterbirds might explain another migratory mystery.

"One of the things that blew us away the very first winter we did Project SNOWstorm was the discovery that young Snowy Owls were spending weeks at a time out in the middle of frozen Lake Erie," says Scott Weidensaul. "We were thinking, 'What are these birds eating out there?' But then we started watching the NOAA satellite images of the Great Lakes and realized that Lake Erie might be 98.5 percent frozen, but it was 1.5 percent open water. As the winds were shifting, these huge plates of ice were breaking up, creating open areas of water that were a lot like polynyas, in which there were grebes, loons, ducks, and gulls. And that's what these birds were feeding on. So you have to wonder, were the owls practicing out on Lake Erie a lifestyle that they might follow if they decide to winter up in the Arctic, out on the polynyas?"

Why do Snowy Owls use so many different migratory strategies? Why do they tend to move to new breeding sites from year to year? We know that their breeding success depends on their finding areas with a very high density of lemmings and that the lemming population fluctuates widely in different geographic areas over time. Most of these owls can't afford to be faithful to a breeding site in the summertime like so many other kinds of migratory birds can, because they can't necessarily count on sufficient numbers of lemmings from one year to the next. They must go where the prey is, so they actively search for lemmings across thousands of miles of Arctic terrain. They're able to do so because they're such powerful fliers.

"What's interesting to me," says Holt, "is how quickly the owls can assess lemming numbers and respond. They just don't miss."

We don't know how a highly nomadic bird like the Snowy Owl finds locations with plentiful prey. "That's one of the big mysteries, how Snowy Owls can track the lemmings in time and space," says Weidensaul. "If you've got an owl that's breeding in the central Canadian Arctic one year and in Greenland the next year, how did it know to go to Greenland? We thought maybe as the birds are flying north, they're stopping and prospecting for lemmings, but looking at our tracking data, that doesn't appear to be so."

According to Holt, there's only one way to find out. "We have all these Snowy Owls flying around with satellite transmitters on them. Now we need to verify, on the ground, what's happening when they move. We need to go to the new breeding areas and to the old sites. We need to go back to the area they vacated and determine whether they left because lemming populations were low, then go to the new sites and ask, How many Snowy Owls are actually breeding there? What's going on with the lemming populations? That's the only way we're going to figure out how Snowy Owls track lemmings. If we just make inferences from satellite maps, we're not going to be able to do this."

The questions abound, but one thing seems clear: these owls know what they're doing, and their decision to stay or to go, to tough it out in a place or to seek new ground, is part of a complex calculus we don't yet fully understand.

Seven

An Owl in the Hand

LEARNING FROM CAPTIVE BIRDS

The Greek goddess Athena kept a Little Owl as a sacred pet. In T. H. White's novel *The Sword in the Stone,* Merlyn the magician enlists his companion owl Archimedes to help in the education of the Wart: "Merlyn took the Wart's hand and said kindly, 'You are young, and do not understand these things. But you will learn that owls are the most courteous, single-hearted and faithful creatures living. You must never be familiar, rude or vulgar with them, or make them look ridiculous. Their mother is Athene, the goddess of wisdom, and, although they are often ready to play the buffoon to amuse you, such conduct is the prerogative of the truly wise. No owl can possibly be called Archie.'" In J. K. Rowling's fictional universe, Harry Potter receives both mail and camaraderie from his pet Snowy Owl, Hedwig.

As for real pet owl owners: Teddy Roosevelt possessed a pet barn owl, along with a handful of guinea pigs, two white rabbits, a macaw, a pig, a badger, a pony, a small bear, and a hyena. Florence Nightingale had a pet owl named for Athena. Nightingale's own name may be synonymous with mercy, but there's an ironic edge to her owl story. The

Florence Nightingale with her Little Owl, Athena

British nurse rescued the owlet from some little boys in Athens and hand raised it, carrying it around in her pocket. When war broke out in the Crimea and Nightingale felt called to take her nursing skills to the field, she put Athena in her attic before she left, thinking the bird would feed on mice in Nightingale's absence. But the little owl was too domesticated to hunt for herself and sat waiting for her meals. When Nightingale and her family failed to check on her in the days leading up to Nightingale's departure, Athena died—perhaps of loneliness and certainly of starvation.

James Gordon Bennett Sr., founder of *The New York Herald*, collected live owls. Bennett, the story goes, took his father's yacht to sea in the Civil War, and one night, a hooting owl awoke him just in time to save the boat from running aground—the start of a lifelong fixation on owls. In addition to his pet owls, Bennett collected owl statuary, promoted the preservation of owls in his newspaper, and fashioned a new building for the *Herald* in the 1890s that included a roof lined with dozens of bronze owl figures, complete with eyes that flashed with electric lights in time with nearby clocks tolling the hours. Two of those owls now guard the statue of Minerva in Herald Square in New York City. (Bennett also planned an enormous owl-shaped mausoleum for his final resting place, but that odd monument never came to be.)

In 1946, Pablo Picasso found an unusual muse while at the Musée d'Antibes in France—not his lover, Françoise Gilot (though she was inspiring, too), or the place itself, a moody medieval château perched high above the Mediterranean. No, it was a tiny, injured Little Owl sitting on a rampart. Picasso adopted the owl, bandaged up its broken claw, and kept the bird until it healed, then decided to take it with him back to Paris.

By some accounts, Picasso saw himself in the bird, aware of the owlish quality of his own face. In any case, it was the beginning of a tumultuous relationship that would ripple through the artist's life and work. In her remarkable memoir, *Life with Picasso,* Gilot writes of the owl, "We were very nice to him but he only glared at us. Any time we went into the kitchen, the canaries chirped, the pigeons cooed and the turtledoves laughed but the owl remained stolidly silent or, at best, snorted. He smelled awful and ate nothing but mice."

The owl was grumpy. Picasso was grumpy. And the two would grunt at each other and snap back and forth. Whenever the bird bit Picasso's fingers, the great artist would hurl equally biting verbal abuse back at the bird. "Every time the owl snorted at Pablo," writes Gilot, "he would shout, '*Cochon, Merde,*' and a few other obscenities, just to show the owl that he was even worse mannered than *he* was." Eventually, the owl would let Picasso scratch his head and perch on his finger, "but even so," writes Gilot, "he still looked very unhappy."

He probably was.

In any case, despite the combative nature of their relationship—or perhaps because of it—Picasso was obsessed, and the Little Owl provided a major motif in his work for the next two decades, appearing often in his ceramics and in drawings and etchings from the period.

Apart from the president, the nurse, the newspaperman, and the artist, not many celebrities have kept owls as pets. And for good reason.

Like most licensed care providers for owls, Laura Erickson will go on at length about the difficulties of keeping a real owl—and also the joys. Erickson is a former editor at the Cornell Lab of Ornithology and host of the radio program *For the Birds*. For seventeen years, she stewarded an Eastern Screech Owl named Archimedes (after Merlyn's bird) and worked with him as an education ambassador, taking him to schools, bird groups, and symposia to teach people about the realities of being an owl.

A family had found Archimedes in their backyard on the ground beneath his nest, in very bad shape. "He was completely unfeathered," Erickson says. "On one side of his tiny body, he had puncture wounds, and on the other side, he was all abraded and scabby." It turned out the owl was suffering from a blood parasite, so he needed a lot of care. The family took him to a rehabilitation clinic, where he was nursed back to health. But because he ended up imprinting on humans, he couldn't be released into the wild. He was later transferred to Erickson, who had a license to possess a screech owl for education. "That's how I was lucky enough to get him," she says.

Feeding Archimedes was a nighttime ritual. She would take a mouse out of the freezer at the start of *The Daily Show*, and by the end of *The Colbert Report*, it would be thawed, she says. "Sometimes he

Archimedes the Eastern Screech Owl at feeding time

would bite it in half and just eat the head and upper torso, saving the back half to eat the next day. Sometimes he would eat the whole thing at one sitting, swallowing it headfirst and then letting it work its way down. You could watch how he would ratchet it, with the lower bill and the upper bill, and the tail would always be the last part to go down, like spaghetti." Each day, around six to eight hours after eating, he would spit out a pellet. To exercise him, she ran him around the neighborhood at night. She would put little jesses or straps around his legs, attach them to a leash, and start running. He would take off and fly right above her head.

Erickson considered her time with Archimedes a sacred privilege. When she was writing in her office late at night with the little owl in the room, she loved calling back and forth with him. Sometimes those exchanges lasted a half hour or more. "There's a joke that we humans are never a cat's owner—they see us as staff," she writes. "I don't think Archimedes saw me as staff, more like his mate."

All of the owl stewards I encountered had a similar attitude toward their birds and a similar personal relationship with them, extraordinary and hard won. Since 2016, Jim Duncan has had the "pleasure of looking after and worrying about" a Long-eared Owl named Rusty, an ambassador bird for programs at Discover Owls, an educational organization Duncan founded and directs. "One of the fascinating things about looking after a live bird in captivity, becoming intimate with them," he says, "is experiencing some of the more subtle communications these birds might make in the wild. Rusty tells me when she's mad. She growls at me. And the intensity of the growl is probably proportionate to the emotion she's feeling. But when she's really feeling cuddly and nice, she'll give me little tiny vocalizations, *huh-huh-huh*, and those can get intense, too." In the spring, Rusty displays to Duncan, indicating that she wants him to nest with her. "But I can't. I'm already banded," he jokes, pointing to his wedding ring. "It's complicated."

Duncan has also noted Rusty's keen observation of family members and her personal preferences. "I have two sons, and she can tell which son is coming down the kitchen stairs without seeing him just by the pattern of how he walks or descends the stairs," he says. "For some reason, she just doesn't like one of our sons, even though he's never done anything wrong to her. She can tell right away when he's coming downstairs, and she starts snapping her bill and puffing up before he even comes around the corner into the kitchen. On the other hand, she takes a real shine to our other son. It's quite fascinating." Duncan looked after another Long-eared Owl, Nemo, for ten years. "And he couldn't have been more different. He had this very relaxed, chill attitude about things. These birds are individuals. Like us, they have their quirks, their likes and dislikes."

Karla Bloem says her Great Horned Owl, Alice, has always been a good judge of character. "I dated some guys that Alice really didn't like, and it was a really dicey thing. As soon as they got near Alice, she would go into attack mode. But she was okay with my current husband from the get-go. It was jaw-dropping. Like he could go into her room, her territory, and hoot to her, and she was perfectly fine with it. She would just bow her head and hoot right back at him. And I was like, 'Oh my gosh, this is a miracle.'"

The emotional bond that forms between human and owl can be profound. When master falconer Rodney Stotts, who works with a Eurasian Eagle Owl and other raptors, was asked what it felt like to develop a connection with such glorious birds, he said, "There hasn't been a word created for that feeling. That bird lands on your hand, and you're just going to cry from the emotions that run through you. You realize how small we are and how small our problems are."

One August day in 2017, Erickson lost Archimedes. He seemed sort of sluggish that morning, she says. He called a little, but he didn't seem his normal self. She went over to him and preened his face a couple of times, and he leaned into the preening, like he always did. She left the room for lunch, and when she returned, he had toppled off his perch onto the floor, dead.

"I like to imagine him free," she wrote in a tribute to him, "flying under the moon or alighting on a massive limb of a big, old elm, or peeking out of a Pileated Woodpecker hole in an aspen, or whinnying under the stars somewhere out there in the universe. I like to imagine that he's shaken off his obligation to teach people things they should already understand about the ways of owls and the value of nature—shaken it off the way he'd shake out his feathers after preening, a spray of feather dust forming a sparkling halo around him."

Arguably the most famous pet owl of the twenty-first century, real or imagined, is Hedwig, Harry Potter's Snowy Owl. Small wonder J. K. Rowling picked a Snowy for the starring role. The big white bird is the world's most distinctive owl, with that gleaming white plumage. In Rowling's universe, Hedwig and other owls straddle the spheres of the magical and the Muggle as part of an owl post service, ferrying birthday greetings, packages, and even a Nimbus 2000, a wizard's broomy vehicle, with efficiency and apparent ease. But they also serve as the wizards' affectionate friends. When Harry needs it most, Hedwig provides warm, comforting companionship.

After the release of all seven of the Harry Potter books and the films based on them, Erickson—aka "Professor McGonagowl"—made herself something of an expert on the owl appearances in the movies and even started a blog called *The Owls of Harry Potter*, inviting questions from readers.

In the films, Hedwig is female, Erickson told me. But the owls playing her were all male, and different males filled the role: Gizmo, Kasper, Oops, Swoops, Oh Oh, Elmo, and Bandit. Why male owls? "For one thing, females are bigger and heavier, harder for a young actor to handle, and they have powerful talons," says Erickson. Daniel Radcliffe, who plays Harry in the films, wore a thick leather guard on his arm beneath his big black cloak. Talons that can nail a merganser through thick feathers can do a number on human limbs. Also, she says, "the pure white plumage of older males compared to females looks more striking against a wizard's black robes." And why so many substitutes or stand-ins? Because apparently owl actors also need days off. Moreover, she says, owls are individuals "and as tricky to train as cats. One might be excellent at doing directed flybys, another might be much calmer sitting in a tiny cage on a movie set, and another better at interacting gently with a little boy while sitting on his arm."

When one of her readers asked Erickson whether a Snowy Owl could really carry a Nimbus 2000, Erickson went to her neighborhood hardware store and weighed a corn broom—just over a pound, about half as heavy as a Snowshoe Hare, which Snowy Owls regularly carry to their chicks. So yes, a real Snowy Owl could carry a Nimbus 2000. But the Snowy Owl in this scene simply flew through, and then the filmmakers digitally added the broom. The owls often dangle their legs when taking off or landing, says Erickson, so in the scene, the owl's legs are already in position to be carrying something.

Years after creating Hedwig, J. K. Rowling apologized to her audience for some elementary errors in her depiction of Snowy Owls. Unlike Hedwig, Snowies are mostly diurnal, she explained (although as noted, they also hunt in crepuscular hours). They don't hoot as Hedwig hoots to comfort Harry or express her approval (their vocalizations are more like a bark). They don't nibble on bacon rinds as Hedwig does when she delivers the mail at breakfast time. Perhaps most important, they do not make good pets. "Much like making Horcruxes, this practise belongs in fiction," she wrote.

"Don't even think about taking an owl as a pet," says Erickson. "They are wild birds requiring a wild life. In cages they simply cannot do all the things their bodies were designed for and their spirits require." Moreover, they're hard to care for. "They eat whole rodents or other whole animals, which must be fresh, and their droppings are messy and smelly, requiring frequent cleanup." It's a massive commitment, a way of life. But this hasn't stopped people from trying. The books and movies spurred a tidal wave of interest, and in parts of the world, boosted the owl pet trade. "In the era of Harry Potter," she says, "I'd get at least a couple of emails a week asking how one could get a pet owl."

In most countries, keeping an owl without a special permit is illegal. The United States forbids people to have owls as pets. Only if you're trained and licensed can you keep an owl for rehabilitation or for educational purposes, or, in some states, for falconry. But even in these situations, the licensed person doesn't own the bird, just "possesses" it and is legally responsible for its care. The US Fish and Wildlife

Service has stewardship of the birds and can recall them at any time if the licensed person or institution is not meeting permit conditions.

In the United Kingdom, it's legal to buy a pet owl if the bird is captive bred. You don't need a license or any credentials. Moreover, owls bred in captivity can be sold without any regulation, and it's a lucrative trade. A Snowy Owl can bring in about £250. In the wake of Harry Potter, so many people bought pet owls in the UK, only to dump them after realizing the cost and complexity of caring for them, that a special animal sanctuary opened to adopt the unwanted birds.

The "Harry Potter effect" was perhaps even more pronounced in the Far East. In 2017, researchers with the Oxford Wildlife Trade Research Group looked at the abundance of owls in the bird markets of Indonesia before and after Rowling's world of wizardry appeared on the scene. In Java and Bali, hundreds of species of wild-caught birds have been offered up for sale as pets for generations, write the researchers. This is because birds in general symbolize accomplishment in Java. "Traditionally, in order to reach a full life, a man had to have a house (*Wisma*), a wife (*Wanita*), a horse (*Turangga*), a dagger (*Curiga*) and a bird (*Kukila*)," they write, "with the horse representing ease of communication and movement within society, the dagger representing status and power, and the bird all of nature, as well as the need for a hobby in a well-balanced life." In the past, owls—known as Burung Hantu, "ghost birds"—were little in demand, but after the release of the Harry Potter books in Indonesia in the early 2000s, their popularity boomed. In one market in the 1980s, 150,000 birds of 65 species were for sale, and not a single owl. These days the markets sell hundreds of owls, called "Burung Harry Potter." In the larger bird markets in Jakarta and Bandung, up to 60 owls are for sale at any given time, including scops owls, barn owls, bay owls, wood owls, eagle owls, and fish owls. In US dollars in 2020, a scops owl would go for as little as $6, a Barred Eagle Owl for $90. People learn about how to care for them from online forums, blogs, websites, and Facebook groups with more than 35,000 members. Pet interest groups meet regularly in parks to show off their owls and exchange information.

Japan is the largest global importer of owls, accounting for more than 90 percent of the thousands of owl imports. This may have to do with "Kawaii," Japan's "cute culture," which lately has focused on owls, with their big eyes and other humanlike facial characteristics. Species ranging from scops owls to white-faced owls to Great Horned Owls are frequently put on display in "owl cafés," successors to the country's trendy cat cafés. While Japanese law specifies that only captive-bred birds are allowed to be displayed in these cafés, traders often get around this with forged documents and the illegal recycling of old permits. Customers are allowed to pet the owls as if they were cats. On its website, the highly popular Akiba Fukurou café promises customers a relaxing experience with its twenty resident owls.

Relaxing for humans, maybe, but not for the owls.

Most of what we know about the psychology of owls, their complicated motivations and nuanced behaviors, we've learned from captive owls like Archimedes, birds that have been injured or human imprinted and therefore can't be released into the wild. One thing these birds have taught us: there's nothing blunt about an owl—except, perhaps, for the way they eat their prey.

Consider Papa G'Ho, a captive Great Horned Owl at the Wildlife Center of Virginia.

At the moment, Papa G'Ho is clacking his beak fiercely and puffing up his feathers so he looks almost twice his size. He's perched in the far corner of his flight run, a good forty feet from where I stand. But when you're in the same enclosure with a male Great Horned Owl displeased with your presence, it's easy to forget that you're three times his size.

"He doesn't like people," Amanda Nicholson tells me. She would know. As a wildlife trainer and educator here at the center, she is something of an expert at reading owl behavior.

Papa G'Ho was admitted to the center in 2001 with damage to his wings and feet, probably the result of being hit by a truck or car. He recovered, but the vets realized he had sustained a permanent injury, the loss of his alula, a small bone at the tip of the wing that's critical for all kinds of flight maneuvers. The bone itself has feathers embedded in the soft tissue surrounding it, and when it moves, it elevates or depresses the feathers, changing the airflow over the wing to control different flight movements, such as deceleration and landing. "It's one of those amazing owl things," says Nicholson. "For a bone so tiny, it makes all the difference in the world in terms of how silently an owl flies and how much it has to flap." The injury meant that Papa G'Ho could not be released back into the wild and would have to remain at the center. There he has a reputation for being fierce, growly, uncooperative—and at the same time, among the most doting, tolerant, and tender surrogate fathers around.

I've come to the center to learn what brings owls here, what kinds of threats they face, the injuries they suffer, and what the highly skilled vets and trainers have learned by working with these birds up close.

The center is a busy place. In the spring, it's full of baby animals, dozens of bear cubs and deer fawns, eaglets and box turtles. Since its founding in 1982, it has provided health care to some 90,000 wild animals, including thousands of owls. It's a teaching hospital with a state-of-the-art veterinary clinic, a diagnostics laboratory, a surgical suite with a big window where students can watch the repair of fractures and eye injuries, a radiology room complete with X-ray machines, and one basic but very necessary waiting room. There's an ICU for very young animals that need an incubator or hand feeding every day, as well as an isolation room for contagious patients. A large kitchen is stocked with nuts and produce donated by local grocery stores.

The whole complex of buildings is tucked into the woods adjacent to a national forest very much like the habitat these animals come from, and its patients recover in an array of spacious outdoor enclosures—aviaries and large flight pens. The autumn day I was there, one aviary was chirping, cawing, and cooing with Carolina Wrens, American Crows, and Mourning Doves. The newest patients just checking in were an Eastern Meadowlark that had struck a plate glass window and a Canada Goose with a broken leg.

Most patients are ferried to the center by a rescuer or picked up by a volunteer transporter. But an unusual delivery occurred one April when a brood of five owlets arrived as stowaways. A family of barn owls nesting in a semitruck full of hay parked in Arizona went on an unexpected journey across the country, and the nestlings were discovered, sans parents, only after the truck reached Virginia. The chief veterinarian at the center, Karra Pierce, was there to receive them. "They were dehydrated and thin but otherwise healthy," she recalls. "That night was the first time I heard barn owls screaming. It was eerie. I thought I was under attack."

The owl patients here are primarily Great Horned, Barred, and Eastern Screech Owls, most of them suffering wing or eye damage, likely from a vehicle strike. "But each case is a mystery," Nicholson tells me. "The patients can't tell us what has happened. And it's not like with pet cats and dogs, where the owners can relate the story. We have to piece it together." The vet team uses sophisticated diagnostic techniques, examining the owl's body, doing blood work, and inspecting fecal samples. The stories that emerge from the vets' careful sleuthing reveal the threats these owls face on a daily basis. They also become teaching tools.

Papa G'Ho, for example, is a poster boy for what's known as the "apple core" message.

If you're like me, you've thought nothing of tossing the core of a Gala or Granny Smith from your car window, assuming that it's biodegradable and may even make a meal for some wild creature along the roadside. Well, it often does—and sometimes to devastating effect.

Imagine a mouse or a vole creeping out of the woods to discover that apple core at the edge of the road and settling down to nibble it in situ. An owl in a nearby tree swoops in for an easy snack at just the wrong moment and gets hit by a passing car. "A mouse on the side of the road is an easy target for the owl, and it's completely focused on its prey," says Pierce. "It has no idea a car is coming."

A few years ago, one Eastern Screech Owl might have met its end this way, but for a fortunate fluke: The car had its window open. The owl flew straight in the window, bounced around a little, and finally landed on the passenger seat. "That was the world's luckiest owl," says Pierce. "If it had hit that car literally half a second earlier, it would have hit the windshield instead." The driver of the car drove its unexpected passenger straight to the center.

Of the eighty or so owls admitted to the center each year, most have probably been hit by a car while searching the roadside for food. Vehicle hits most often cause ocular trauma—damage to the eye—or wing or leg breaks. "These are things we can generally fix," says Pierce. The vets create micro-bandages, set bones, and make miniature splints for their feathered patients. They restore essential flight feathers on the wing and tail through "imping," a remarkable technique in use since the thirteenth century in birds used for falconry. "We clip the damaged feather near the base, leaving a hollow keratin sheath," explains Pierce. A matching feather molted from a donor bird is cut to length and inserted into the sheath with a tiny keratin dowel and then superglued to keep it in place. "If all goes according to plan," she says, "the imped feather serves as a good replacement for the original feather until the bird molts it naturally, dropping the imped feather and growing a new one."

Other injuries are internal and require more sleuthing, like those suffered by the young Eastern Screech Owl a homeowner found tucked in a bush, bruised and lethargic. When the vets at the center examined the fledgling, they noticed it was bleeding from tiny puncture wounds, and its blood would not clot—a classic sign of rodenticide poisoning. A blood test confirmed the diagnosis. The rodenticides people use to kill rats contain toxic compounds that inhibit blood clotting, which eventually leads to severe internal bleeding and death—not just for the rat itself but for the raptor that feeds on it. Fortunately, this little owl had ingested only a small amount of the poison and slowly made a full recovery. After five months of care at the center, it was returned to the wild.

Barry the Barred Owl wasn't so lucky. For many months until her death in 2021, Barry charmed visitors to New York's Central Park. She was famous for her daytime displays of preening and stretching in the hemlock tree in the Ramble where she roosted and for splashing about in a little stream and then shaking her feathers, unfazed by onlookers who traversed her home territory. "Barry chose to live in Central Park because it suited her own needs, but she was a wonderful gift to New Yorkers," says photographer David Lei. "It was a joy to see her stare down the squirrels who boldly confronted her, to watch her hunt after dark, and to hear the sound of her distinctive 'Who cooks for you? Who cooks for you all?' call sound over the Ramble." One day when Barry was flying low in the park, she collided with a maintenance vehicle and died from the impact. A necropsy showed that she carried a potentially lethal level of rat poison, which might have impaired her flying and her judgment.

Sometimes an owl will show up cold and unresponsive, in what Pierce calls an owl "coma." Was it a head injury? Something ingested?

Again, blood work can yield answers. One young Eastern Screech Owl found sitting quietly on the ground at a city park in Charlottesville had no apparent external injuries, but an X-ray revealed a skull fracture, and a blood test showed a significant amount of lead coursing through the owl's veins.

"Something to consider is did this owl sustain neuro or head trauma because it was lead intoxicated?" says Pierce. "It's kind of like drunk driving, right? You're at higher risk for accidents if you're doing things when you're impaired."

Testing owls for lead poisoning is a relatively new thing. Lead analysis used to be routine only in eagles and vultures, scavengers that were likely to feed on the gut piles that hunters leave behind, laced with lead fragments from bullets. But there's a growing concern that owls are also ingesting lead from spent ammunition, perhaps by eating a squirrel that was shot but not killed and still carries bullet fragments. Pierce recently conducted a study looking at blood lead levels in four kinds of owls over the past ten years. The results were disturbing. Thirty percent of Barred Owls tested positive for lead, 30 percent of Eastern Screech Owls, and 19 percent of Great Horned Owls. "Unfortunately, lead is everywhere," says Pierce, in opossums, squirrels, even box turtles.

The vet team started the young lead-poisoned screech owl from the park on a course of therapy to scrub the lead from his system. After four or five treatments, he recovered, but the lead poisoning had lasting effects. Lead is a neurotoxin and can cause permanent neurological problems. The little owl had a "raging head tilt," says Nicholson. "He held his head almost upside down." So Pierce made him a tiny neck brace to help him keep his head upright. "But he still flies on an angle and wouldn't be able to fend for himself," she says. "That owl is not going back to the wild." Instead, he's now working with Nicholson to become an ambassador bird.

The goal of the center is "to treat to release"—to restore patients to health and return them to the wild. But release is only one of three possible outcomes. Another is euthanasia for owls that can't be released because of a permanent injury that compromises their balance or vision or hearing, or joint or bone issues that might later cause pain. This is a big bucket, unfortunately, but vets and rehabbers consider euthanasia a kind of gift for an owl that's suffering. (In fact, the term means "good death.") The third outcome is a life of captivity, sometimes with a role like the one Papa G'Ho plays, as a surrogate parent or as an education or ambassador bird.

Papa G'Ho's job is to help baby owls get to that first, best outcome. He may be cranky with people, but by all accounts, he's an awesome role model for Great Horned owlets.

Baby owls need this. When an owl chick hatches, it doesn't instinctively know what it is. Imprinting on a member of its own species helps a baby bird learn and interpret species-specific behaviors and vocalizations so it can choose appropriate mates later in life. If an owlet imprints on a human, the owl will never be owl enough to survive in the wild. (That's what happened to Karla Bloem's Alice, Jim Duncan's Rusty, and very likely Florence Nightingale's poor Athena.) It's easy for an owl to make the mistake. Our faces look owllike to them, just as their faces look humanlike to us. As Laura Erickson put it, "They respond to us the way we respond to them." If a baby owl imprints on humans, it won't fear them. But it won't necessarily be friendly toward people either or welcome their contact. In fact, as Alice demonstrates, human-imprinted owls can be territorial and aggressive with people, just as they would be with other owls.

Rehabilitators have learned all this the hard way. Now they go to extraordinary lengths to avoid human imprinting in the young owls they care for, keeping contact at a minimum, wearing masks and camouflage balaclavas to conceal their faces, and avoiding any kind of communication with the owls. "We try to be invisible to them so they don't get used to humans, and we're completely silent," says Pierce. "We post big signs all around where we have babies—NO TALKING! BABIES INSIDE!

"We really prioritize getting them with conspecifics. Owlets need to see another member of their species to learn, 'Oh, I'm an owl.'"

Here's where Papa G'Ho comes in.

In the spring of 2021, a trio of tiny Great Horned Owl hatchlings were found on the ground in three different areas of Virginia. The birds may have jumped out of their nests, been pushed out by a sibling or pulled out by a predator, or they may just have been fledging, trying to learn to fly. In cases like these, efforts are made to renest young birds, put them back where they came from, reuniting them with their parents. The rescuers will even create a new nest if the original has fallen down—then watch to see if the mother comes back. But when that's not possible, the orphans are brought to the center to be examined and treated if necessary. Then, if they're Great Horned Owls, they're placed with Papa G'Ho to learn essential owl behaviors.

In the company of the big, old owl, the chicks learn to be wary of humans and defend themselves by fluffing their feathers and huffing and hissing and clacking. They learn to socialize with one another. "They're extremely curious and active and fun to watch," says Pierce. "They hop around and tear stuff apart. Papa G'Ho is tolerant and a great exemplar. His very presence shows them how to recognize their own species so that when they're released, they can mate with a member of their own species."

Papa G'Ho also teaches them to be fearful of humans and to use appropriate defensive behaviors. "He raises some fierce children," says Nicholson. His owlet students "are very defensive toward humans, clacky, huffy, and they 'mantle' a lot," hunching, crouching, and spreading their wings over prey to conceal it.

Papa G'Ho the Great Horned Owl with chicks

In the past two decades or so, Papa G'Ho has raised more than fifty owlets. Never has he rejected a chick or been aggressive toward one. So impressive is his fathering that one year he won the coveted Coolest Dad award from *Virginia Living* magazine along with a human father. The human dad won a T-shirt. Papa G'Ho won ten dollars' worth of mice.

After a spring and summer with their role model, young owlets are ready for the next steps in rehabilitation, flight conditioning and mouse school.

"Flight conditioning is really just a fancy way of saying we gently encourage them to fly back and forth in their enclosure," says Pierce. "We send them on a prescribed number of flights, like swimming laps in a pool. We need them to do five good laps before we up them to ten. And we put them through that whole process again and again until they've built up their stamina and their flight is good enough to be competitive in a wild owl population."

The rehabbers also look closely at the owls' flight form. Are their legs in the appropriate position? Are they flying straight? Are they flying silently? "We want to hear nothing when they fly," says Pierce. Noisy flight means failure—the bird will not be released if it can't fly quietly. But if owls can pass their flight test, they go on to mouse school.

Young captive owls attend weekly mouse-school sessions and then must pass a final test to make sure they're able to hunt on their own. For the test, the rehabbers put live mice in a horse trough in an owl's flight enclosure, covering them with leaf litter. To pass, the bird must catch the mice overnight for five nights in a row.

It sounds straightforward. It's not.

Imagine you've been spoon-fed mouse pieces for most of your short life. Suddenly, your food moves. It scurries, it hops, it squeaks, it bites. "Oh, this is different!" You might try to hunt with your beak—which you would think might work but doesn't.

"Typically, we drop off the mice, then check the next day to see if they're gone," says Nicholson. It's good news if they vanish. But how long did it take the bird to catch the mouse? "It may have taken two hours of stomping around in there, trying to pick it up with their beak until they're like, 'Oh, right! I've got to do this with my feet,'" says Nicholson. "I've watched young owls go through this transition. It's such a steep learning curve."

"And you start doubting yourself," Pierce adds. "You're like, 'Well, did the mouse escape? Did a skunk come in and get it?' Anything can happen. So we decided for insurance purposes to put a critter camera in there, to make sure the hunt is going as it should and that the owls are actually getting the mice."

Adult birds with a confirmed catch on camera pass mouse school in a single night. "I'm satisfied with that," says Pierce. "But the babies have to go through more nights to make sure it wasn't like, 'Oh, I can kill white mice but not brown mice' or, 'You know, I had a bad experience with a very feisty mouse last night and I decided I don't want to try that again.' So they go through a longer process than adults." The same is true for any owls with ocular impairment.

One male Eastern Screech Owl was offered live mice for three nights in a row but apparently found the whole business so unexpected and intimidating that he just gave up and didn't eat. In mouse school, you either eat live prey or you go hungry. This owl chose to go hungry.

"And we're like, 'Oh, gosh, what are we going to do?'" says Nicholson. "So we paired him up with a buddy who already knew how to do mouse school. He quickly learned from his peer, and after a few nights was good to go."

On the other hand, some owls at mouse school are overeager students. "We have some very disturbing footage of one Barred Owl that was really gung ho about the whole thing," says Nicholson. "She wouldn't eat dead food very well, so we tried live mice." Normally owls wait to eat until the coast is clear. "It makes you very vulnerable to take your eyes off a potential predator in the room to go hunt and eat your prey, closing your eyes and swallowing it and all that," she says. "But this owl did not even wait for humans to leave the enclosure. The rehabber dropped the mouse in the tub, and the owl just dove in, grabbed it with her talons, transferred it to her beak, and swallowed it whole. You could see the mouse still moving. I thought, 'Oh my God, aren't you even going to kill it?'

"That they all eventually figure it out just blows my mind."

During a Zoom presentation in late February to benefit the center, a Barred Owl named Athena appears onscreen with Nicholson, perching calmly on her gloved hand. Athena is an ambassador bird.

Because they're so charismatic, owls are among the most popular ambassadors in animal educational programs around the world. Sometimes they're trained to simply sit on the gloved hand of a trainer, sometimes to fly in a free-flight show in a room or the outdoors. Either way, they captivate audiences and—the hope is—inspire people to care more about the birds and their environment. Of the three possible outcomes for an owl at the center, this one, becoming an ambassador or education bird, is the smallest bucket, says Nicholson, "but it's the one that has taught us the most about the inner lives of these birds and how to read their behavior."

I met Athena at the center. She had been there since 2012, when she was found on the ground in Richmond, Virginia, weak, uncoordinated, with retinal degeneration in both eyes. The vets thought she might have been affected by West Nile virus. In any case, she was partially blind and couldn't be released. Nicholson says that working with this bird over the years has given her insights into owl behavior she would never have gotten otherwise. She has trained other kinds of raptors, too, and the differences are telling. "I look at one of our hawks or the Peregrine Falcon I work with, and I can easily see them processing in their brains what's happening," she says. "They're easy to read.

"Owls, on the other hand, are subtle in showing what they're thinking, and you have to be in tune with them to read their behaviors. Some falconers I've worked with consider owls the 'dumbest' of raptors and impossible to train for that reason. Now that I've trained owls myself, I'm like, 'Oh, no, you guys are very wrong. Owls are not dumb. You're just not good at reading their behavior.'" Think about it, she says. "We humans are daytime creatures, blundering around in the bright light and generally making a lot of noise. Owls fly by the lowest light and hunt in the dark, silently navigating by the slightest of sounds. And here we are, loud and obnoxious and really bad at reading their behavior."

Athena the Barred Owl

This is perhaps especially true for small owls, such as Northern Saw-whet Owls and Eastern Screech Owls. Nicholson says that small owls were once thought to be easy to train, compliant and cooperative. "But that's because we were misunderstanding their behavior," she says. "There are a lot of people in our field who have liked working with Eastern Screech Owls and other small owls because they're so 'calm.' There's this outdated notion that they're easygoing. In reality, these small owls are terrified. They're on high alert and are trying to keep a low profile. They're holding completely still because they're trying to be invisible, trying not to die. They're saying, 'Here I am, I'm just part of the tree, blending in with the tree.' And we misinterpret that. The owl may be doing what we ask, but inside it's dying."

I think back to handling that female saw-whet at the banding station in Powhatan, how apparently docile she was, how still. She was doing what small owls do in the face of a threat. She was trying to disappear.

Nicholson says that everything she knows about interpreting owls and owl behavior she has learned from Gail Buhl of the Raptor Center at the University of Minnesota, one of the country's leading authorities on training rehabilitated captive owls.

"There's no question that owls teach you to be a better observer of behavior," says Buhl. "The signals are hard to read, but they're so important because they reveal the owl's inner state."

Two young Northern Saw-whet Owls, each blind in one eye, taught her how to understand small owls and how to train them to be ambassador birds. One was female, named Acadia; the other, male, named Me-Owl. "The birds don't know their names," she says, "but humans like to name things. When we name the ambassador owls, we pick a name that has to do with the bird's injury or its natural history or its scientific name so it becomes a touchstone for educational discussion." Acadia is where the Northern Saw-whet Owl was first described in the scientific literature. Me-Owl was caught by a cat, who blinded her eye. ("Cats do a horrible amount of damage to wildlife," she says.)

Those two small owls taught Buhl five big things.

First, when you're training an owl, you need to consider the natural history of that species, which shapes its response to human contact. Where does it live in the world? In cavities, on the ground, in the deep forest, or on the open land? What are its defenses when faced with a predator? "Look at what that owl does in the wild, how it protects itself," she says. Saw-whets stay still and slit their eyes.

People try to train small owls as if they were a big predator like a Red-tailed Hawk or a falcon or even an eagle. "But subtle things make owls different. Yes, small owls eat small mammals, reptiles, birds, insects. But they also get eaten by other things, by larger raptors, snakes, ground predators. Yes, they're predators, but they're also prey."

Second, owls are individuals. They have different personalities and respond to training in different ways. Buhl trained Acadia in just three months. It took a year to train Me-Owl, who—perhaps understandably—was warier.

Third, if there's a choice between safety and food, owls will choose safety. So finding positive reinforcement, making them feel safe enough to take a reward, is challenging. Buhl learned to use those little bristle feathers alongside an owl's beak to help her. "These birds learn early in their lives that they don't see very well up close and that their bristle feathers help tell them where the food is. They think, 'I catch a mouse, I reach down and feel the mouse with my bristle feathers, this is the mouse, this is my toe, and lo and behold, I get the food in my mouth.'" So Buhl learned to touch the bristle feathers, and nearly always the owl would turn toward her and open its beak, and she could pop in a small piece of food as a reward.

The fourth and biggest lesson was this: owls have emotions. This was a major shift in thinking about these birds. "When I first got into the field of ambassador animal training, the idea of owls having emotions was frowned upon as anthropomorphizing," she says. "The attitude was: 'We don't know anything about that, so we're not going to talk about it. We're just going to deal with behaviors.'" But during her training, she realized that owls certainly do have emotions, and those emotions affect their behaviors. "Being fearful of something is going to affect how you respond to it. And we need to recognize that those two things, the emotions and the behavior, are connected."

Finally, the saw-whets taught her what emotions *look* like in a small owl, how fear and distress show up, as well as feelings of comfort, relaxation, of "safety." And, perhaps most important from the perspective of the bird, what triggers these feelings, how the presence of a human, especially a human trainer, affects an owl. When Buhl teaches other owl trainers, she starts with this: What do we (as humans and trainers) look like to owls? And how do owls show the emotions triggered by our presence?

To begin with the obvious, she says, we're huge compared with a small owl. "And we act like predators. We like to look people and other animals straight in the eye. We stare, just as predators do, keeping their eyes on the prize. If I'm staring at the owl, I'm acting like a predator." (Camera lenses, too, are predatory from an owl's perspective.) "We walk directly to owls. That's also predatory behavior. We walk and then stop and ask ourselves, 'Is the bird looking comfortable?,' then we start walking again, slowly, making our voices quieter. All of this resembles creeping or stalking, like a lion or a housecat. Staring at your target, wanting to get as close as possible, creeping before pouncing."

And how do owls respond? Small owls react to this with an acute distress response, also known as a fight-or-flight response. In many birds, the first defense is to fly away, but in small owls, "the flight response is a freeze response because they depend on camouflage to hide," says Buhl. "They think, 'If I sit still, you, the predator—photographer or trainer—will go away because I can depend on my camouflage. That's what I know.'"

Buhl has worked extensively with all kinds of raptors in wildlife rehabilitation. "Rehab elicits pretty extreme defensive behavior in other raptors," she told me. "Red-tailed Hawks get their hackles up and extend their wings. Bald Eagles bite and wing slap. Vultures bite and twist. These are all defensive strategies. Some owls will bite, too, but for most, their go-to defense is to sit still and be quiet. They're stoic about stuff, and people misinterpret that. They're experiencing the same stress as other raptors, but they're internalizing it."

In her talks, Buhl shows two images of Short-eared Owls, photographed in roughly the same position. The one on the left has its eyes open wide and popped out, almost bulging, like little marbles. Its pupils are dilated, its feathers slicked back, held tight to the body, and its facial disk pulled taut. Its feet are "grippy," and it's almost standing on its talons. Its plumicorns are raised. Short-eared owls are classified as non-tufted, says Buhl, but they show their tufts in situations where they feel they're in distress. It's kind of a vertical cringe.

"This is not what you want to see."

The owl on the right has eyes partly opened; its pupils are small, its facial disk relaxed, its feathers uniformly fluffed. Its feet are tucked in, "a sign of relaxation," says Buhl, "like us when we cross our legs or ankles while we're sitting in a chair."

This *is* what a trainer wants to see.

The two owls look like entirely different species (as did those Long-eared Owls in Milan Ružić's photographs from the same tree in Kikinda). The visible distinctions reflect what's happening inside the bird's brain. The owl on the right feels safe and comfortable. The one on the left is in a fearful freeze response.

"Short-eared Owls also depend on camouflage to hide, so they're saying to themselves, 'No matter how much distress I'm feeling, I have to stay very, very still to blend in with my surroundings. That's the response I go to because of my natural history, tightening everything, even pulling my wings in very close, gripping my perch.' It seems counterintuitive because the response of other birds is simply 'I'm out of here, I'm just leaving.' But most owls depend on camouflage first."

Buhl tells her students to read a bird by looking at its eyes, its feathers and tufts, its feet, and its body posture. All these body parts reflect an owl's feelings, its emotions, its thoughts. Then, she says, try not to be a predator. "Try to help the owl realize you're not a threat through your body language, by averting your eyes, holding your body at a forty-five-degree angle away from the bird, even backing into the room, only glancing at the bird, then looking away, dipping your head to peek, using only your peripheral vision to see the entire bird."

What Buhl and Nicholson and other trainers have learned about the psychology of owls certainly helps them improve their training techniques. But it also offers an intimate view of the hidden inner lives of owls, their subtle emotions, sensitivity, and stoicism, which changes the way I experience these birds—and photographs of them.

Since Buhl began working with owls, a major shift in attitude has taken place among experts in the field rooted in an acknowledgment of the emotions of owls and their nuanced expressions. "Now training is based on a greater understanding of a bird's sensibilities and behavior and relies on setting up a relationship of trust," says Buhl, "where the owl gets to choose to engage and participate in an activity."

Nicholson's approach to training Athena is a good example. "She was always reactive to weird noises," says Nicholson, "like crunchy leaves and scraping gravel." She realized these were the noises she would make when she entered Athena's pen and was moving toward the bird to try to get her on a perch to weigh her. Now Nicholson asks Athena to come to the perch on her own, to voluntarily get on the weigh scale. "Some days, she says, 'Nah,'" says Nicholson, "and I say, 'Okay, I'll be back later.' But most of the time she gives a clear sign, 'Yes, I will participate, and I'm okay!'" Training is a conversation between trainer and owl to help it control its own environment. It's about doing things on the bird's terms.

Only a very, very small number of owls become ambassador birds. Buhl and her colleague at the Raptor Center, Lori Arent, have a set of stringent criteria for selecting owls to be ambassador animals. They're equally picky about the educators with whom they place their owls. "These placements are often for life," says Arent. So there's a lot that goes into it. If the birds are well taken care of, they can live well past their normal life span in the wild. "We want to make sure that as they get into their geriatric years, that they have proper medical care and people who can adapt their management," says Arent. "They can develop age-related cataracts and arthritis and other conditions. So we want to make sure that they're going to have good care and good training for that whole time. All of this goes into a decision on both ends to make a good match."

And all of this puts into stark perspective the matter of the average person possessing a pet owl.

When I ask Buhl about this, she's careful to say that the question is bound up in culture, in differing views on wildlife.

"I don't think it's a good idea for all the reasons we've been talking about," she says. "I don't think people understand what they've got. But I'm speaking as someone who was born and raised in Minnesota. I don't want to pass judgment on other perspectives."

"That said, we do struggle with it. We see videos of a Snowy Owl raised with a human family, and the kid is hugging the owl, and I'm watching it going, 'Oh my God, that bird is *this* close to ripping that kid's face off.' I can see this in its body language. And that part drives me crazy. If you don't interpret behaviors correctly—and most people aren't equipped for that—somebody can really get hurt."

Also, the owls suffer, "even owls that are captive bred, human imprinted, or human raised," she says. "I've seen videos of the owl cafés in Japan and watched people getting close to the birds and giddily putting their phones up to them. I'm watching the behavior of the owls, and I'm thinking, this poor animal is dying inside, but people don't see it."

We see what we want to see. "We view them through our own lenses," says Buhl. "We can't help it. But we need to treat them not as mini-humans in feathers, but as their own entity, intrinsically what they are, their own 'nation' on this planet," she says—her words echoing the great naturalist Henry Beston on all wild animals: "They are not brethren, they are not underlings; they are other nations, caught with ourselves in the net of life and time."

This is an important truth. But the human urge to see owls as mirrors of ourselves, symbols and reflections of our beliefs, has been present and powerful for a very long time, perhaps since the beginning of our species.

Eight

Half Bird, Half Spirit

OWLS AND THE HUMAN IMAGINATION

Bone carving of an owl by Moche artists of Peru's north coast (third to sixth century AD)

In December 1994, three cavers led by Jean-Marie Chauvet exploring a remote valley in southern France noticed an air current, an updraft venting from some rocks in a limestone cliff. When they dug away the debris, they were able to squeeze into a passage through the rocks and descend into a massive cave. On its walls were paintings of horses, lions, aurochs, and woolly rhinoceroses that looked so fresh, so newly drawn, that the cavers at first doubted their authenticity.

The paintings were in fact among the oldest examples of cave art ever discovered, more than twice as old as the paintings at Lascaux. They looked fresh because they had been preserved, frozen in time when a massive rockslide sealed off the cave tens of thousands of years ago. Among the drawings was an owl, the first known representation of a bird.

One day some 36,000 years ago, an ancient member of our species walked into the Chauvet Cave, down through the darkness to its far reaches, with only a torch or grease lamp for light, stood before an overhanging rock, and scraped clean the surface. Then, with a finger, carved in the soft yellow surface of the rock a fine image of an owl almost a foot and a half high, complete with ear tufts and vertical lines representing wing feathers. Judging from the ear tufts, it was probably a Eurasian Eagle Owl or a Long-eared Owl, drawn with its head turned around to face the viewer; its body is shown from the back, suggesting its ability to turn its head 180 degrees.

"One sees the Owl whilst returning from the deepest depths of the cave," reads the official description—"is it looking back down there, with its inhumane ability to see in the dark?"

Humans never lived in the cave, says eminent French prehistorian Jean Clottes, the first specialist to investigate Chauvet after its discovery. It was used only for painting and maybe for ritual and worship. The owl, like the other animals, was drawn not out of casual interest or for aesthetic pleasure but to invoke spirits, he argues, to ask for their help.

Fast-forward some 15,000 to 20,000 years. People in the Magdalenian culture in the same region were using Snowy Owls for mysterious purposes. Southern Europe was different then, and so were its owl species. The cold climate of the Upper Pleistocene probably favored the presence of Snowy Owls, and Magdalenian hunter-gatherers in the region brought their remains into caves and rock shelters. At one site in France, Combe-Saunière, scientists found the bones of close to a hundred Snowy Owl wings—not the carcasses, just the wings. At another, Saint-Germain-la-Rivière in southwestern France, archaeologists discovered the remains of 22 male and female adult Snowies, some with decorative carvings on the bones. These finds suggest that 20,000 years ago, hunter-gatherers were using not only the meat of the owls but also their talons, long bones, and feathers. "It is plausible that hunter-gatherers took advantage of the winter irruptions to catch large numbers of them," writes Véronique Laroulandie, who studies archaeozoology in the region. The complex, meticulous way they processed the owls "clearly does not relate to just basic food needs," she says. "Even the meat could have been consumed not strictly for alimentary purposes, but for medicinal or even magical reasons." Some 200 miles to the south of Saint-Germain-la-Rivière, at the cave of the Trois Frères in the Ariège, is an "owl gallery" with a pair of owls—probably Snowies—depicted in profile and face-to-face. Between them is a young owl. Laroulandie is intrigued by the ambiguity of the images. Do they depict owls or owlish-looking humans? Is it just we who perceive this, looking back through the ages, she asks, "or was it intended by the artists to evoke a relationship that connected them to this owl?"

How long have we been obsessed with owls?

Forever it seems. We evolved in their presence; lived for tens of thousands of years elbow to wing in the same woods, open lands, caves, and rock shelters; came into our own self-awareness surrounded by them; and wove them into our stories and our art. We know their shape the way we know the shape of a snake. They're in our myths, our dreams, our DNA.

Today owls are vibrant players in cultural stories from every continent. David Johnson is on a mission to fathom the full range of human attitudes toward owls across time and space—and to understand their origins. For his Owls in Myth and Culture project, he assembled more than a thousand reports of owl lore and stories from around the world and through the ages, and his collection is growing. Thumb through these reports, and you'll find that owls are indeed imbued with all kinds of meanings—they're creators, healers, guides, and guardians, as well as fearful, devilish presences, harbingers of doom and death.

Among the Ainu people of northern Japan, owls are considered manifestations of the "Owl God," or kotan-koro-kamuy, the protector and advocate of humans. The Blakiston's Fish Owl of Hokkaido is revered as a god who brings fish to the river. According to Ken MacIntyre and Barb Dodson, the authors of a report on the Noongar (also called "Nyungar") culture of Australia, "two night birds, the mopoke (*Wau-oo*) and the white owl (*Beenar*), were said to have created all the Bibbulmun (Nyungar) people and given them their social structure during the cold, dark, formless period of the *Nyitting* (Dreaming)." The Yahgan people, who dwell on the islands south of Tierra del Fuego, identify the barn owl as sirra, the wise grandmother, who generates new streams and rivers to provide the fresh water that allows their people to survive. Today the Yahgan still associate the sounds the owl makes in flight with the susurrations of streams.

If the gods took their wings from owls, so did the devils. In many cultures, owls are leagued with witches and malevolent spirits, with sorcery and death. "Human fear of owls is common in every part of the world," says Johnson. "The form it takes may vary from culture to culture, but there are patterns that pop up, consistent beliefs and narratives in different versions."

According to Heimo Mikkola, an expert on owls and owl lore, it's a common belief in several African countries that "if an owl crosses the road while you are traveling you can be sure to get some misfortune." If an owl perches on top of your house in Zambia, you will receive bad news or there will soon be a funeral, and if one sits in your farm fields, your crops won't grow. Similarly, among Algonquian people of North America, "persons about to die hear the owl call their name." This is true for the Kamba of Kenya, too, write ethnobiologists Mercy Njeri Muiruri and Patrick Maundu of the National Museums of Kenya. The hoot of an owl perched on or near a house is perceived to mean "someone will be sick or will die," usually within fifteen days. "To counter this, the Kamba place a piece of a clay pot on the branch where the owl was perching." For the Ch'ol Maya in Chiapas, the Mottled Owl (known as kuj) is considered a sabedor ("knower"), write Kerry Hull and Rob Fergus, who have studied bird beliefs in Mesoamerica. "When it cries 'jukuku jukuku,' it is a sign that someone will die." Similarly, the K'iche' Maya regard owls as heralds of sickness and death, and they are extremely unwelcome anywhere near homes.

Why are owls such powerful symbols? Why are they considered spirit beings and gods and heralds of good fortune in some cultures, and bearers of bad luck, harbingers of doom, and malevolence on wings in others?

Our species has found meaning in the presence of birds ever since we were—well—human, says Felice Wyndham, an ethnoecologist affiliated with Oxford University. "People everywhere and throughout history have seen birds as messengers and signs. We are omen-minded. We're always looking for hints about what's coming up, for predictions and premonitions, and we turn to the natural world for signs. It seems to be a human universal."

Wyndham studies the domain of birds "that tell people things."

Ancient Egyptian limestone relief (circa 2030–1640 BC)

Birds move between earth and sky, day and night, so who better to turn to for clues? There's even a name for it, *avimancy*, from the Latin for bird and the Greek for prophecy.

Recently, Wyndham and her colleague Karen Park collected reports of birds as signs from cultures around the world. "We started it as a lark," she told me. "In our review of ethnobiology accounts, we kept coming across references to birds as signs. Then, just listening to people talking in cafés and on the bus, I started to notice, oh my God, *everybody* is aware of birds as signs and are still really using that in their daily lives."

There was the anecdote about a finch's visit to Vermont senator Bernie Sanders. One Friday afternoon during the 2016 election primaries, Sanders was giving a speech at a campaign rally in Portland, Oregon, when a finch landed on his podium and perched there, just to the left of his trademark A FUTURE TO BELIEVE IN sign. The crowd went wild, and so did the internet: "Birdies for Bernie." "That bird was feeling the Bern." Among Sanders supporters, it was a fortuitous sign. "The birds have spoken."

And then there was the story Wyndham was told during her fieldwork in the Ayoreo communities of Paraguay. One day a hummingbird flew past a community leader—considered "a portent of very good news." The next day, the leader won $1,000 in the lottery.

Wyndham and Park found close to 500 reports from 123 ethnolinguistic groups across six continents—from the Ojibwa to the Mbuti, from the Maori to the Welsh to the Basques. In all these cultures, people see birds as harbingers—omens of luck or ill fortune—and as envoys, often from the world of spirits or the divine. Because the reports were most often written in English, Wyndham and Park knew they might be missing the nuance and subtlety that would show up in original languages, the different meanings of bird signs. Still, it was clear that people find birds especially good for "knowing things with," as Wyndham puts it.

All variety of birds appeared in the ethnographic accounts, herons and woodpeckers, doves and sparrows. But when the pair counted the incidence of different birds, one order popped to the top: owls. In 57 of the 123 ethnolinguistic groups, owls were considered meaningful signs.

When my ninety-six-year-old aunt was near her end but still cogent, she told her daughters that after she died, she would send them a message, a sign that she had reached the other side. Later, my cousin wrote to me to say that the night her mother died, an owl "kept hooting nearby. Maybe it was just a coincidence," she said, but in the forty-one years she had been visiting her mother, she had never heard an owl hoot anywhere near the house.

In many parts of the world, owls were—and still are—viewed as messengers bringing news from this world or others, from the past or the future. Among the Yup'ik of Alaska and the Russian Far East, Great Horned Owls are said to speak the Yup'ik language and drop feathers from their wings as they convey their messages to people about the days ahead. In Athenian Greece, owls were emblems of a mántis, the Greek word for diviner, prophet, or seer. Mánteis "would put a little statue of an owl on their table, a kind of 'calling card' to declare their role as prophet," says Wyndham. "It was like hanging out a shingle. Everyone knew that if you put out an owl, that was a service you gave."

Sometimes an owl calling near at hand is considered simply a natural sign, a warning of impending bad weather or the announcement of the presence of a plant or animal.

The Iroquois of North America interpret an owl calling nearby as a signal that they should gather wood against imminent cold and snow. "Something similar was also present in field cases in southern Chile," says Tomás Ibarra, a human ecologist at Pontificia Universidad Católica in Chile. One individual who was interviewed said that when a barn owl cried during the night in the hot season, "it would dawn with dense mist, and it did, it was like a barometer that we had."

In the Tenejapa region of Chiapas in southern Mexico, the Mottled Owl is known as mutil balam, "bird of the jaguar," or mutil coh (choj), "bird of the cougar," and its appearance suggests the big cats are close. For the Apurinã people of Brazil, the call of the caburé owl, the Ferruginous Pygmy Owl, indicates that peccaries are nearby. If the owl is silent, hunters will remain at home.

These "folkloric" views of owls as natural signs may seem quaint, but they're often rooted in real natural history observations. In many cases, the "reading" of owls is related to a sophisticated understanding of ecological relationships, and the birds may well be providing essential information about the environment.

The ubiquity of owls as signs may have something to do with their presence on nearly every continent, says Wyndham. But it goes much deeper than this. "It's their humanlike visage that makes owls such potent symbols," she says, "along with that whole grab bag of features": their nocturnal nature, their strange swiveling neck that some people find deeply unnerving, their peculiar voices that emanate from the night, and their sensory powers beyond human capabilities.

I think of writer J. A. Baker's description of a Tawny Owl: "The helmeted face was pale white, ascetic, half-human, bitter and withdrawn. The eyes were dark, intense, baleful. This helmeted effect was grotesque, as if some lost and shrunken knight had withered to an owl."

"For me personally, their spookiness has a lot to do with their silent flight," says Wyndham, "how they just appear out of the dark without advance notice. There's a whole literature in evolutionary psychology about how people understand the way the world should work. Living things aren't supposed to be able to walk through walls, or to appear and disappear without integral motion. Owls seem to break those rules and defy expectations. That's part of why we put them into a supernatural category."

All of these owl qualities generate an "uncanny" feeling, says Tomás Ibarra, who studies the role of owls in human societies. Owls are like us but also not like us. "They have this human quality, rounded head, flat face, big eyes, which has led some societies to call owls 'the human-headed bird,'" he says. And yet most species live in a world where we humans feel uncomfortable—the night. Moreover, their eerie screeches and hoots are unlike the musical songs and calls of most birds. Also, they are wild creatures that appear in domestic settings, which is unsettling. "People feel connected to the owls but at the same time don't like seeing them close to the house," explains Ibarra. "There's this feeling, 'you don't belong here.'" It's the unfamiliar in a familiar context, and also the reverse, the familiar in an unfamiliar context. "When we see this," he says, "we experience what we call 'chicken skin' in Spanish," or goose bumps.

Ibarra has loved owls since he was a child. Like David Johnson, he had a transformative owl encounter when he was just six. He was living in an agricultural area of Santiago, Chile—now densely populated, but at the time rural, with fields, streams, and groves of big walnut trees. One dark night his father took him on a walk through the orchards to a grove of the largest walnut trees. "My father shone a light to the top of a huge walnut tree," he recalls. "A big barn owl there opened his wings and flew off. I've never forgotten that." A few years later, the family moved to a more isolated area in the Andes mountains, and one night while camping, the teenage Ibarra heard the deep, booming call of a "tucúquere," a Lesser Horned Owl. That was it, he said. "I went to southern Chile to study owls for my master's and met the Rufous-legged Owl, my favorite species. I'm crazy about *Strix*; they're like witches."

At first Ibarra concentrated on ecology, specifically whether owls were good indicators of forest diversity. Because they're top predators, owls have a special correlation with high levels of biodiversity, and they're known to be sensitive to environmental pollution, he says. "But I've always been interested in culture, too, so now I'm trying to bridge ecology and anthropology." Ibarra focuses on the concept of biocultural memory, the idea that human cultural practices and beliefs grow out of observations of the natural world, and that our memories are a result of these interactions between cultural traits and our environment.

"Owls have a lot to say about biocultural memory," he says. In Chile, for instance, they're not only woven deeply into the folklore, but crop up in traditional sayings and proverbs that people use on a daily basis. "Búho que come, no muere." (The owl who eats will not die.) If a sick person eats, he or she will not die. Or "Lechuza vieja no entra en cueva de zorro." (An old barn owl does not enter a fox den.) That is, old age and wisdom foster sound decisions. "Mata el chuncho." Kill the chuncho—the local name for an Austral Pygmy Owl—and you'll avoid bad luck. That owls are so present in Chilean proverbs and expressions suggests how integrated they are into daily life and imagination, says Ibarra. And yet they have this strange, mysterious aspect. It's this mingling of the familiar and the unfamiliar that gives owls their uncanny quality, he says. "They're like us in their appearance, but they're so unlike us in every other way. And that's what we find so troubling and exciting about them."

To explore what people believe about owls in a wide range of different cultures, David Johnson and his project team conducted more than 6,000 interviews in 26 countries, from Argentina to Turkey, Belize to Tanzania, China to Ukraine, Pakistan to Gambia, Ethiopia, Mexico, and Mongolia. The interviewers surveyed some 200 people from each region and asked them a range of questions: Do you consider owls spirits or real birds? How much do you know about owls, their ecology, habitats, nesting, and diet? How many species can you identify? What stories about owls have you heard? What does your culture believe about owls? Where possible, Johnson enlisted native speakers to try to capture the subtleties of thinking and beliefs. "We wanted to examine and explore these questions in an unbiased and scientifically focused way," he says.

The responses to the question "What do you believe about owls?" landed all over the map. People considered owls powerful spirits, wise, creator beings, scary, dangerous, helpful for medicine, bad omens, bearers of bad luck—an especially prominent belief in Belize. Most Argentinians said owls are just birds. Iranians considered them wise but also bad omens and bearers of misfortune. In northwestern China, the majority of people said they believed that owls were helpful or brought good luck. "But when we got down to the part of the survey that asked 'Have you or someone you know killed an owl?,'" says Johnson, "one hundred twenty of the two hundred responders said yes."

Of the twenty-six countries surveyed, twelve had a positive view of owls, eight were neutral, and nine had a negative view. In six countries, people indicated that underlying cultural beliefs were shifting in a positive direction, but in some countries, there had been no change in beliefs in the past one hundred fifty to two hundred years.

One of the indicators for changing views in societies, Johnson found, is the quantity of owl-related merchandise in the marketplace. "If owl-related things are readily available, that's a good indicator of broader acceptance in a society," he says. In Brazil, Turkey, and other countries with some positive momentum, there were owl-themed backpacks, clothing, jewelry, clocks, figurines, even beer. But after two years of searching the marketplaces of Belize, where owls are still considered bad luck, the team found only two items: a child's backpack and a slate carving.

"If an owl comes to your house and calls, it means that someone will get sick or die." This view of owls, or very slight variations of it, turned out to be the most prevalent perspective around the world, says Johnson. "It's not that owls cause death; they're just the bearers of bad tidings." In any case, their presence is highly inauspicious.

In many cultures, owls are considered partners in witchcraft or witches themselves morphed into bird form. In the Olancho region of Honduras, the lechuza, or barn owl, is "without doubt the most hated and feared of Olancho's birds," ethno-ornithologist Mark Bonta told me. Some people claim that one type of barn owl, the lechuza de sangre, sucks the blood of infants. "The latter is an evil spirit, la lechuza," he says, "a witch in avian form." The idea of owls as "witch birds" is prevalent in parts of Africa, too, in Malawi and among the Yoruba of Nigeria, Benin, and Togo. When researchers in Zambia asked children to write what they knew about owls from stories passed down in their families, witchery was a common theme, especially owls as a means of transport by people who practice witchcraft. "Witches are said to use owls to get from one place to another during the night," wrote one child. "So that is why people fear them . . . because they think they might be bewitched by the witchcraft that is in the owl."

Owls as socially controlling beings is another common theme. Parents using owls to control or discipline their children shows up from culture to culture, says Johnson. "It's 'the white owl will get you' kind of stuff, and it's more common than you might think."

It's not hard to understand how a half-glimpsed barn owl, ghostly white, with its strange funereal night cries and habit of haunting vacant buildings, might give rise to the notion of a birdlike incarnation of a demon or spirit being.

Nor is it surprising that owls would be linked with menace or threat. "Nocturnal birds in general are often associated with heightened danger," says Felice Wyndham. "After all, like other primates, we're more vulnerable to attack in the dark." Moreover, on occasion, owls have been known to attack people when they feel threatened or while trying to protect their nests. In 2015, a Barred Owl defending its territory in Salem, Oregon, attacked joggers so often that it earned the nickname Owlcapone. As I write this, a Barred Owl has been startling the Brookwood Hills neighborhood of Atlanta, swooping down on people at dawn or dusk. The Georgia Department of Natural Resources has recorded eighty strikes, including some victims who have been hit multiple times. But this is extremely rare, and in the case of the Atlanta bird, a spokesperson from the DNR suggested that the owl may be human imprinted and was trying to capture people's attention to seek food. One Barred Owl was even implicated in the murder of a woman in 2001 (though most experts agreed that the wounds did not look like those an owl might inflict, and the woman's husband was eventually found guilty of the crime).

Jonathan Haw, who runs an owl education project in South Africa, speculates that owls may have become linked with imminent death because of our own circadian rhythms. Cardiac events such as stroke, heart attack, and sudden cardiac death most often occur in the final hours of the night. He points to research suggesting that 3:00 to 4:00 a.m. is the time when we struggle to maintain our body temperature and when adrenaline and anti-inflammatory hormones are at their lowest, causing airways to narrow. "In rural areas without hospitals and where a lot of people die at home, that's generally the time that we die, just before dawn," says Haw. "And at that time, people are more likely to bump into an owl sitting on a roof or hear one calling in the dark. The owl might have been sitting on the roof forever, looking for rats, but suddenly people kind of put that whole thing together and this is where it goes."

Marco Mastrorilli, an owl expert in Italy, has another theory as to why some people may yoke owls with death. Among some Catholics in Italy, it has been a tradition to lay out the dead in front of the house and surround the corpse with candles. "The candles draw moths, and the moths draw owls," he says.

In some parts of India, owls are known as Madhe Pakhru, meaning "corpse bird." Researchers from the Ela Foundation in India, an organization devoted to owl conservation, investigated one aspect of the link between owls and death in Maharashtra, a populous state in western India. They visited fifty-seven cemeteries and found that owls were present at fifty-three of them, including seven different owl species, from Spotted Owlets to Indian Eagle Owls. Owls are drawn to the cemeteries because of the presence of big trees and the lack of

Plate from Goya's Los caprichos*: "The sleep of reason produces monsters" (El sueño de la razon produce monstruos), 1799*

human disturbance, the team surmised. "In some instances food offerings are kept at the cemeteries during the performance of last rites and this attracts rodents," which, in turn, attracts owls. People become aware of the owls because the time of burial is at dawn or dusk, when the birds are actively calling—hence the "haunting" association between owls and death.

Many of these overtones, both sunny and dark, crop up not only in myths and stories but in visual images of owls through the ages, dating back to those ancient cave paintings. In the art world, too, owls speak of wisdom and luck, of misfortune and malevolence.

Several of Picasso's etchings and prints feature owls inspired by his pet Little Owl. "They're mostly R-rated," says Robyn Fleming, a research librarian at the Metropolitan Museum of Art, "scenes of orgies, with the owl just sitting there in the midst of the revels": *Bacchanal with an Owl*; *Nocturnal Dance with an Owl*; *Prostitute, Sorceress, and Traveler in Clogs*, with an owl at the center gazing provocatively at the viewer.

The birds in Picasso's bawdy prints are only a few of the owl images that abound in the collections of the Met. Fleming is on a quest to find them all, carved in ivory and stone, hammered in gold, woven into tapestries, brushed onto canvas and ceramics, and drawn and painted on the pages of manuscripts.

Some of the images she has found are realistic and show evidence of keen natural history observation—the mobbing of owls by small birds, for instance. Others are symbolic, loaded with superstition or spirituality and revealing attitudes toward owls ranging from deep admiration to derision, from humor to fear.

Fleming began her search in 2017 because of a mild interest in what she calls the "catness" of owls. As a kitten, her cat had looked like a Great Horned Owl. "I didn't know how I felt about that," she says. "I thought owls were creepy. But since then, I've really embraced them." After a quick spin through the Egyptian galleries, where owls popped up often because they're a hieroglyph for the letter *M*, she realized the birds were everywhere in the museum's collection. "So the way I got into owls is kind of weird but finding and researching images from the collection has been a blast." She even got a tattoo of one of them.

On Friday evenings when the Met is open late and very quiet, she wandered the galleries, hunting for owls, eventually creating an "owl map" of the entire museum. Some owls were obvious and easy to find, like the nineteenth-century gold brooch in the form of an owl head or the Ptolemaic relief plaque with the face of an owl from 400 to 30 BC.

Others required sleuthing and sharp eyes. Fleming discovered a whimsical owl on the back of a fifteenth-century Italian majolica apothecary jar, hidden from the viewer. Only the human profile on the front was visible. She pored through the Met's online collection for references to owls not on display in the museum and, with the help of curatorial staff, tracked them down in storage and study rooms. She's especially proud of discovering a huge Russian woodcut of a Eurasian Eagle Owl that took her breath away and a truly minuscule and "ridiculously adorable" little owl tucked on a branch in the background of a sixteenth-century Dutch painting by Joachim Patinir, *The Penitence of Saint Jerome*. She began posting the images on Instagram around the time other Met staff members were launching a menagerie of accounts—"Cats at the Met," "Dogs at the Met," even "Butts at the Met."

Now Fleming is known at the museum as the "owl lady." The 550 images she has found so far reveal the wildly weird, wonderful, idiosyncratic—and very long-standing—human fascination with owls. The oldest depictions in the collection are those hieroglyphs from the Egyptian Old Kingdom, dating from 2575 BC. Among the most recent is a striking collage by the first-grade winner of a 2019 art contest the museum holds every year for the New York City public schools. "The winners get to have their art displayed in the museum for a few months," says Fleming. "Among the contestants, there's often an owl.

"I think I'm pretty good at this point knowing which genres of art *might* contain an owl," she says. Art (like market merchandise) can reflect cultural acceptance or rejection of owls. "I know to look for them in the decorative borders of European prints and decorative arts and in certain mythological or biblical scenes, and where not to bother looking too hard—in most contemporary art, sub-Saharan or Oceanic art. But you never know!"

The reverence for owls among ancient Greeks shows up in thick, heavy silver coins featuring a helmeted profile of the goddess Athena

Tetradrachm, an ancient Greek silver coin, minted in Athens 480–449 BC

on one side and a Little Owl on the other, a symbol of wisdom. Athenian "Owls," as they were known, were minted in Athens beginning around 510 BC and continued in circulation for more than 400 years.

The artwork of the Moche people, who thrived on Peru's North Coast centuries before the rise of the Incas, around AD 200 to 850, is full of owl iconography. The Moche admired owls as nocturnal hunters capable of finding prey in utter darkness. They considered them warrior creatures, and the birds appear in prominent spots on their helmets and weaponry. A remarkable Moche bone carving showing an owl with ear tufts and feathers forming a wide circle around its eyes probably served as a thumb rest for a deadly weapon, an atlatl, or throwing stick. The Moche also viewed owls as messengers who ferried defeated warriors to the world of the dead, just as they would carry their prey to the nest. One gorgeous gold Moche nose ornament in the shape of an owl head, made in the sixth century, was found in the tomb of a young lord, tucked inside his mouth.

On the flip side, in paintings and etchings from northern Europe beginning in the Middle Ages, owls show up as partners in human crimes of greed, gluttony, depravity. In *The Temptation of Saint An-*

thony, attributed to sixteenth-century painter Pieter Huys, a wide-eyed owl looks like it has witnessed something truly fearful. According to the Met's museum notes, it's held by "a nude temptress who represents the demon disguised as a queen." In the centuries after, owls are often depicted as malevolent night birds perched on the shoulder of a drunk (in keeping with the old Dutch saying "drunk as an owl") and overseeing other scenes of foolishness, debauchery, and vulgar behavior. Hieronymus Bosch, known for his macabre and frightening depictions of hell and his fantastic imagery, almost always put at least one realistic owl in his paintings, often watching over the scene. And, of course, there's Picasso's series of owls in compromised positions.

Among Fleming's favorites are elegant turquoise owl beads carved by the Zuni people of her home state, New Mexico, and a tournament shield in wood and burlap from Germany, circa 1500, showing a realistic owl surrounded by a motto that reads in translation: "Though I am hated by all birds, I nevertheless rather enjoy that." If that isn't an owly attitude, I don't know what is.

But it's the naturalistic images from Japan that stole Fleming's heart and adorn her body. "With so few strokes, the Japanese artists somehow capture the whole essence of the bird," she says. "They're very observant of nature and seem to have a particular appreciation of birds."

Three owl images on Japanese scroll paintings became contenders for a tattoo—a reward for getting through the rough waters of the past few years, she says. "Once my job was stable, Trump had gone, vaccines were coming out, and everything was looking positive, I went for my treat." Three hours later, on her left forearm (placed prominently, where she can see it), she had a five-inch tattoo of a solitary owl on a pine branch. Exquisitely detailed, delineated entirely in black ink, the "brushwork" layered in tones to convey the texture of feathers, the owl

Owl Mocked by Small Birds, *silk album leaf by Kawanabe Kyōsai, circa 1887*

appears exactly as it does in a scroll painting from the early seventeenth century, alert, eyes wide open, with its head turned slightly to gaze provocatively at the observer, as if to say, "I'm here. Listen up."

Poring through Robyn Fleming's images and David Johnson's reports, I'm tempted to divvy human attitudes toward owls into neat categories: superstitious versus those grounded in empirical belief; realistic versus fanciful portrayals; perceptions of owls as portents of evil and ill tidings on the one hand and as good omens, revered birds of wisdom, on the other.

But it's not as simple as that.

Framing human beliefs about owls in a one-dimensional way risks pigeonholing them, says Felice Wyndham. "What fascinates me is that for so many people there's not a clear division. It's all mixed up. It's imagination *and* empirical observation, poetic significance *and* practical application, reverence *and* fear. With owls, you can't easily separate them out."

In many parts of the world, owls are hitched to sorcery and sickness but also credited for their role in rodent control. In Mesoamerican cultures, they're often seen as bad omens and harbingers of death, but they also occur as parts of personal names, and for hunters, they're an auspicious sign, signaling the proximity of good game. While owls are traditionally revered as creator beings in Noongar culture, owls and owllike night birds are collectively called *mopok warra,* meaning bad or dangerous. In some Mapuche oral stories, owls symbolize fear of the unknown and the untamable, but in other stories, they're beneficent and wise.

The point is, few animals have such an immensely ambivalent and complex relationship with humankind. From folklore to art, they've been revered and reviled, deemed sage and stupid, coupled with destructive witchcraft and with healing. Sometimes they symbolize two opposing things at once. And sometimes they're just . . . birds.

"Owls are powerful for so many people," says Wyndham. "They tap into our deepest feelings. The real picture we must hold is of a bird order that arcs through our full moral compass, from spiritual being to reviled omen to cherished symbol of wisdom."

All of this is important to understand, in part to grasp the rich tapestry of human attitudes toward owls, and in part because there are consequences. Human views are not simple. Nor are the implications for owl conservation.

Nine

What an Owl Knows

HOW WISE ARE OWLS?

Barred Owl

It's late summer in the southern Appalachian Mountains. A narrow trail winds between oaks and hickories tinged yellow and, higher up, through spruces, pines, and firs. No owls in sight, but I know they're here. These woods are full of them. Barred, Great Horned, Eastern Screech, and now—I realize—Northern Saw-whets.

Owls have changed the way I see this landscape, the snags and felled trees not as debris but as nurseries and ramps for branching owlets, the scrubby gullies not as ecological wastelands but as hideaways for roosting owls. I think I spot a screech owl nestled in a snag, but it's only a stubby broken limb doing a credible owl imitation. Ha! Turnabout is fair play.

I stand and listen. It's daytime. The owls are quiet. They see me but stay unseen, so well hidden they escape my eye, even though they may be yards away.

Writing this book has grown my wonder at these birds. Owls see what we don't see. Hear what we don't hear. Invite us to notice sights and sounds that might otherwise go unnoticed. With their quiet, subtle presence and cryptic coloring, they point to the value of not standing out in the world but fitting into it. For owls, invisibility is a defense or disguise; for us, it's a privilege, one that—if we're lucky—may yield an owl sighting.

Owls teach us what we can learn from an animal just by listening to it. They show us how distinctive they are as individuals, idiosyncratic, with as much personality as we have, and with a full range of feelings and emotions often expressed in deeply understated ways. They tell us that to get at their truths, we need to understand them over time. It's not enough to seek quick glimpses. We think we know something about them, and then—*poof!*—they dispel our theories, offering up bent or broken rules and unexpected qualities.

Here's another example of the newly discovered enigmatic abilities of owls. In 2021, scientists at the Israel Institute of Technology discovered that barn owls build mental maps of their surroundings in a part of the brain called the "hippocampus"—while they're on the wing.

The hippocampus is a brain structure that helps animals navigate their environment. It underlies our human sense of direction, sense of place, and our ability to organize our memories of life experiences. The structure contains "place cells"—neurons that selectively fire at particular spots along a route. Walk from your kitchen to your living room, and the different single neurons, or place cells, will fire, encoding a map of the route in your hippocampus. When you travel that route again, the same place cells fire in the same order. In this way, a sort of structural memory of place is built inside the brain. Scientists study this cognitive map by recording single neurons in animals exploring space.

We used to think place cells and mental maps were unique to mammals, including humans, rodents, and bats. Birds didn't have them. But then scientists found place cells firing in the brain of a songbird, a Tufted Titmouse, as it walked along the ground. And now they've been spotted in a flying owl. This makes good sense if you think of an owl's need for familiarity with its territory. The barn owl "strongly relies on memory to navigate to strategic standing posts and to its roost at night," write the researchers. "We speculate that the owl's ability to hunt and navigate in nearly-complete darkness is made possible, in part, by an exceptional hippocampal-based spatial memory."

So Milan Ružić was right about those Long-eared Owls in Kikinda that seem to know exactly where they're going. They likely do have a mental map in their heads—yet another owl superpower.

The discoveries by owl researchers deepen our understanding and our awe. But there's so much we still don't know. I think of how little we know of the numbers and movements of so many owl species. Or the complex ways they communicate, feed, and sleep. Or how an owl hears its own flight and adjusts it while hunting, or the way Snowy Owls find the big populations of lemmings that sustain them from one breeding season to the next. Or how high up in the sky migrating saw-whets fly, or how pygmy owls can whiz straight out of the cavity with no flight practice. Or, for that matter, how smart owls are.

A Great Gray Owl would seem the quintessential "wise old owl," with its enormous head and bright eyes, its body inclined forward on its perch at an angle of forty-five degrees, as if in eager curiosity or interest. Is the expression rooted in any truth?

Great Gray Owl

The "old" part is still a question. Although owls in the wild can live up to twenty-five years, most have a much shorter life span. Big owl species live longer than smaller ones. But we still don't know the details. And the harder problem is the "wise" part. What have owls taught us about their intelligence? How wise *are* they? Is their wisdom only in our imaginations? Or is their intelligence a rich, unexplored area for future research?

We used to think all birds were simple-minded, flying automatons, driven solely by instinct, and that their brains were so small and primitive they were capable of only the simplest mental processes.

An owl's brain, like most bird brains, would easily fit inside a nut—a fact that gave rise to the derogatory term *birdbrain*. But we've known for some time that brain size is not the only—or even the main—indicator of intelligence. And the truth is, most owls have relatively large brains for their body size, just as we humans do. Scientists recently proposed the origin of their big brains: parental provisioning, parents feeding their young during a critical period of their development. When a group of animals called the "core land birds" (songbirds, parrots, and owls) arose some sixty-five million years ago, the theory goes, they brought with them something extraordinary: altricial young, chicks that are hatched in an undeveloped state, requiring parental care. With that trait came extensive parental provisioning of those immature young, which resulted in notably large brains in some bird lineages, including parrots, corvids (ravens, crows, and jays), and owls.

Size aside, bird brains have also gotten a rap because of a perceived difference in their architecture—an apparent lack of a layered cerebral cortex like ours—reinforcing that once-derisive view. The neurons in the cortex-like part of bird brains (called the "pallium") are arranged in little bulblike clusters, like a head of garlic, whereas our neurons are arranged in layers, like a lasagna. We thought that an animal needed a layered cortex to be intelligent. But new research shows that the bird pallium is in fact organized a lot more like the mammalian cortex than we first imagined.

Moreover, scientists have discovered that what really matters in the intelligent brain—however it may be organized—is the density of nerve cells, or neurons. And while the brains of birds may be small overall, it turns out that in many species, they're densely packed with neurons, *small* neurons. This gives the brains of some birds, such as parrots and corvids, more information-processing units than most mammalian brains and the same cognitive capabilities as monkeys, and even great apes. The more neurons there are in a bird's pallium, regardless of its brain or body size, the more capable it is of complex cognition and behavior.

Also critical to intelligence is the connections between neurons—how they're networked, wired together. And in this way, bird brains are not so different from our brains, with some very similar neural pathways. For example, to learn their songs, songbirds use neural pathways that are similar to those we use to learn to speak. Crows use the same neural circuits we use to recognize human faces.

More and more, the pillars of difference between our brains and those of birds are toppling. The latest to go is the capacity for consciousness. A 2020 study of Carrion Crows suggests that bird brains possess the neural foundations of consciousness. "The underpinnings are there whenever there is a pallium," says neuroscientist Suzana Herculano-Houzel.

All of this is humbling and suggests we still have a lot to learn about bird intelligence.

But owls? The knock on owls is that most of the cortex-like part of their brains is dedicated to vision and hearing, some 75 percent, in fact, which supposedly leaves only a quarter for other purposes.

Not long ago, scientists conducted a classic test of intelligence on Great Gray Owls. The so-called string-pulling paradigm is widely used to test problem-solving skills in mammals and birds. The task requires animals to understand that pulling on a string moves a food reward to within reach. Crows and ravens peg the test easily. The experiment showed that Great Grays presented with a single baited string failed to comprehend the physics underlying the relationship between the objects—that is, they didn't grasp that pulling on the string would move the food toward them.

But really, is this a fair test of an owl's intelligence? As Gail Buhl remarks, "It's kind of like telling a rabbit, fish, or antelope that in order to pass an intelligence test, they have to climb a tree."

Owls may not be smart in the ways crows are smart, in the ways *we* are smart, devising technical solutions to physical problems or comprehending the physics underlying object relationships. But this may only point to the limitations of our own definitions and measures of intelligence.

Little Owls are a symbol of wisdom, companion to Athena, the ancient Greek goddess of wisdom. Pavel Linhart points out that there may be something to this. The Little Owls that Linhart studies recognize people, distinguishing between the farmers they're used to seeing several times a day and the researchers who sometimes band them, check their nest boxes, or observe them through binoculars, and behaving differently around the two types of people. With the farmers, they're relaxed and don't take off as quickly as they do with the ornithologists. "These owls are also very curious and investigate their environment," he says, which can make them vulnerable to certain traps around human settlements—vertical pipes, ventilation tubes, hay blowers, chimneys, etc.—they can get in but can't get out. (So curiosity can also kill an owl.)

Ask ornithologist Rob Bierregaard whether owls are smart, and he'll tell you a story about wild Barred Owls. He trains the wild owls to come to a whistle so that he can tag them with a GPS tracker or retrieve the device. "I'll put a mouse out on the lawn, and when they come down to catch it, I'll whistle," he explains. "Then I'll put out another mouse and whistle, another mouse and whistle. After three mice, they'll come when I whistle." The owls learn this in a day, and it never takes longer than three sessions to get a bird completely trained. "I've had birds that were waiting for me the day after I trained them," he says. "So my IQ assessment is based on how quickly they learn that a whistle means free mice."

Bierregaard remembers one owl with an impressive memory for the training. "We called him Houdini because he got out of every trap that we could put up," he recalls. "Three or four years after I first lured Houdini with mice, I went back to his woods to look for him. I whistled, and he came in. It had been *years* since I'd been in his woods, and he remembered that whistle!

"So you put all that together and I think you can assess for yourself how intelligent owls are."

Also, he says, owls are playful, especially young owls. Scientists suspect that play depends on cognition, and that species with relatively bigger brains tend to play more. Play is not easy to recognize or define in the animal world, but evolutionary psychologists have come up with a set of criteria for identifying it. Play is activity that's exaggerated, awkward and inappropriate, nonfunctional, and repetitive. It's generally spontaneous, intentional, pleasurable, and rewarding, and initiated only when an animal is relaxed. "We have videos of owls in the nest pouncing on feathers and jumping in the nest," says Bierregaard. "No apparent purpose to this. Just playfulness."

Bierregaard thinks playfulness in Barred Owls may also explain some of their strikes on joggers and bikers wandering around cities and suburbs. It typically happens in late summer, early fall, he says. "We'll get a rash of people reporting they've been hit by owls. And I think that's just young owls being playful. I talked to a rehabber who had a couple of young owls that were orphans, and he raised them free flying around his house. He reported that by late summer, early fall, the young owls were flying out and about and just whacking people that were riding by on their bikes or joggers, sort of playing tag with them. Of course, if you're playing tag with an owl that has very sharp talons, it's not much fun when you're tagged 'it.'"

Gail Buhl notes that adult owls in captivity will engage with enrichment items left around in their aviaries or pens. They may "foot" toilet paper tubes or egg cartons, tote them around and rip them apart—regardless of whether there is food hidden within. "Many of the small owls will carry around plastic toy insects at night," she says. "They never try to consume them, but in the morning, we find the toys in all kinds of places—sometimes cached with leftover food or cached by themselves or just left on another perch."

Karla Bloem says that the captive barn owl living at the International Owl Center also plays quite a bit at night. "She has stuffed animals, and at night she flies and pounces on them all over the place," she says.

You could argue that pouncing and jumping and carrying around plastic insects is just hunting practice. But Bierregaard points to a video on YouTube of a barn owl in Spain playing with a black cat, the two animals sitting side by side and then suddenly swatting, swooping, and pouncing at each other, then just as suddenly, rejoining to nuzzle. "That just blew me away," he says, "because of all the owls, I would think that barn owls would be the least playful. But these animals are clearly playing."

Jim Duncan sometimes plays with his Long-eared Owl, Rusty, and his Great Gray, named Oska, when he feeds them. He lays out Easter egg cups with a little string attached and hides their food, a dead mouse, under one of the cups. Rusty doesn't seem to get the game, "and if the mouse is invisible under the cup, it's like, 'Oh, the mouse is gone.'" But even if the mouse is completely hidden, Oska flies over and gets to work on lifting the right cup.

Duncan has observed other apparently clever behavior of nesting Great Grays in the wild. After the young start to thermoregulate or to fly, the female may leave the family group and go off to hunt on her own to feed herself and replace the large amount of weight she has lost during the breeding process. (Most of the food the male is bringing in goes into raising the young as quickly as possible to maximize their survival.) But she hunts only in areas outside of the male's hunting territory, so she's not taking food away from her young. At one nest, just before the young had fledged, the male was killed by a car. Instead of taking off, the female stayed with her chick and fed it. "Then what she did was fascinating," says Duncan. "She slowly worked her way over to another family group where a male was feeding his young and the female had already taken off, and she let her partly orphaned chick become part of the family group for that adjacent family. Now her chick was being fed along with the male's other two chicks, so she could take off and hunt to feed herself."

When I asked Gail Buhl and Lori Arent about the intelligence of owls, Buhl told me that her view of this has changed radically in the past decade or so.

"We used to talk about how owls weren't intelligent," she says. "They just had all these great adaptations for survival and didn't really need to have a lot of smarts. They were acting pretty much only from instinct rather than from learning. But I've made a complete one-eighty from this, totally changed my mind. Owls certainly hatch with a lot of the tools they need, but they learn throughout their whole lives. Their survival depends on it.

"There's a continuum of what we call intelligence in the animal world," she says, "and among owl species, there's also a continuum—or just differences." Arent points out that young Great Horned Owls spend six months with their parents learning survival skills, "and if somewhere along the line they get displaced, they starve," she says. "Then you have other species like Barred Owls, which spend maybe six weeks with their parents—if that—learning the same types of skills, and yet they seem to pick them up more quickly.

"But in general, owls are highly intelligent," says Arent. "I think that we have a preconceived notion of what intelligence is. And we can only relate to animals with this preconceived understanding. Owls have a different mindset. They have a different learning style—put it that way. And that's just something humans haven't mastered yet."

Owls, then, are inviting us to find new definitions of intelligence in animals. I think of the way a Flammulated Owl removes the stinger from a scorpion before feeding it to its young. Or how a young Long-eared Owl in Kikinda learns from its peers that the people milling about in the central square are not a threat. Or how Northern Pygmy Owls have "figured out" that synchronous hatching is the way to go to successfully reproduce. Or how, if snags are in short supply, a Great Gray Owl will show its flexibility and seek the stick nest of a Red-tailed Hawk a hundred feet up in a pine tree.

It's true that a large part of owl brains is dedicated to scanning and to sensing in their dark environment. But doesn't that give them their own breed of genius, an amazing capacity to navigate their night world with a skill and deftness we can only imagine?

I think also of a fascinating new study of Burrowing Owls showing what happens when these owls colonize urban areas. While urbanization is robbing good habitat from many species of owls, some, like Barred Owls, screech owls, and Burrowing Owls are successfully colonizing urban areas. Studies show that owls that feed on insects are drawn to the streetlights in cities and will nest nearby to skim the easy meals. In a 2020 study, a group of researchers at the Max Planck Institute of Ornithology made a fascinating discovery: they compared the genomes of urban and rural Burrowing Owls in South America and found that the urban birds, which had started colonizing cities just a few decades ago, had evolved variants in ninety-eight different genes important for cognitive function—genes that could well play an important role in adapting to city environments.

This study points to something hopeful in owls, what biologists call "plasticity" in animal cognition and behavior. Owl populations have within them variation, adaptability, and resourcefulness that allow at least some individuals to respond flexibly to environmental change.

Consider the Long-eared Owls that settle in those huge urban roosts in Serbia. "The birds that are better adapted to human activities are the ones that will pass on their genes," says Milan Ružić. "In this way, these roosts may be shaping the species. Because of human pressures on the natural world, habitat change and climate change, evolution is happening much more quickly than it normally would. It's a time when species must either adapt or die. Maybe what we're witnessing now with these massive urban roosts is not going to be great in the long run. Maybe the owls would survive better if they had never had contact with us. But we don't know because we're in the middle of the process."

I think also of the vagrant Snowy Owls that end up in surprisingly southerly places. This may be accidental, the result of inexperience or a failure in navigating. But for some migratory birds, going "off route" may be a way of charting an escape from a depleted prey base or human-induced climate change or habitat destruction. Instead of staying put in unlivable places and facing potential disaster, a few pioneers scope out new places. This could be true for owls, too. The capacity to make flexible decisions over a lifetime, to learn, shift strategies, explore, and adapt—in the case of urban Burrowing Owls, in a remarkably short period of time—certainly qualifies as a kind of intelligence. And it may be some cause for optimism in the face of the catastrophic changes occurring in the world.

It seems to me that owls are showing us how birds can *embody* intelligence, in their eyes and ears, their cryptic coloring and flight, their memory and hunting skill, their flexibility, nuance, creativity, and discernment.

They are also showing us how animals can offer services in ways we never imagined—economic, ecological, cultural.

Farmers and grape growers in places ranging from California to Serbia, the Netherlands, and India profit from the presence of barn owls, which keep rat and mice populations in check. In Chile, Argentina, and other countries, a variety of owl species—Rufous-legged,

Lesser Horned, Austral Pygmy, and barn owls—hold down the population of hantavirus-bearing rice rats, protecting human health.

As apex predators, wolves of the sky, owls serve an important ecological role, preventing rodents and other prey from crowding out other species, and maintaining balance and preserving the integrity of ecosystems. They show us the interconnectedness of living things. Where would owls be without woodpeckers to drill their cavities and the hawks and magpies to build their nests? And where would nesting Red-breasted Geese be without Snowy Owls to protect them?

What? Snowy Owls protecting a prey species? I know, it's contrary to what one might think. But because of their aggressive nest defense, some owls offer protection against predators for other bird species nesting nearby. Snowy Owls don't tolerate skuas or Arctic Foxes near their nests, and studies show that Red-breasted Geese reproduce more successfully when they nest close to the nests of the big, aggressive owls. Songbirds nesting under the "predator protection" of Ural Owls in Finland enjoy a similar benefit. The owls chase away nest predators while defending their own nests, and also eat other predators in their territory, such as crows and weasels. The protective effect radiates several hundred yards from the nest. What a lovely upsetting of the traditional view of predator-prey relationships.

Owls can act as natural ambassadors of habitat conservation. "People really respond to them," says Denver Holt, at least in some parts of the world. "If we're working to conserve grasslands, tundra, desert, or forest, and there's an owl species living there and relying on that habitat, the owl can really bring home the conservation message."

Owls can also offer surprising glimpses into the ecological past. It's an idea raised in Anthony Doerr's beautiful novel *Cloud Cuckoo Land*. In the sanitized virtual world he has been hired to create, the book's main character surreptitiously undermines the system by slipping in the truth in the form of little owls, "owl graffiti, an owl-shaped drinking fountain, a bicyclist in a tuxedo with an owl mask," writes Doerr. "Find one, touch it, and you peel back the sanitized, polished imagery to reveal the truth beneath"—the calamities, drought, famine, and suffering of the real past.

There is a deep hunch here. Owls may be mirrors of our souls, but they're also windows into what life was like long ago.

When Australian ecologist Rohan Bilney stumbled on knee-deep caches of owl pellets in the caves and rock ledges of Gippsland, Australia, he knew he had found a gold mine of information. The pellets belonged to the Sooty Owl—a particularly enigmatic species that's good at keeping secrets—and they dated back hundreds, even thousands, of years. Their contents revealed a picture of rapid environmental change in Australia that had never been fully understood. What was the mammal community like before European colonization, before the changes wrought by clearing, burning, grazing, exotic herbivores, logging, and the introduction of invasive predators?

"We had such a poor understanding of this," says Bilney. "The scarcity of historical information has made it hard to estimate the magnitude of environmental change, and we've often gotten it wrong." This is especially true for how changes affected forest mammals in Australia, many of which are small and nocturnal, and this lack of knowledge has distorted the perception of the current condition of ecosystems and hindered appropriate management action.

In the owl pellets, Bilney identified the bones and other remains of more than 7,500 individual mammals, many now extinct in the region, revealing enormous changes in the makeup of small-mammal populations. Just before European settlement, Sooty Owls in Gippsland preyed regularly on at least 28 mammal species. Now their diet in this area consists of only 10 species. Scientists knew that Australia has had the greatest mammal extinction of any country in the world, yet the extent of the declines and devastation brought by European settlement had been significantly underestimated. We have Sooty Owls to thank for this new insight.

Owls offer another sort of service, too.

When Priscilla Esclarski was interviewing people for David Johnson's Owls in Myth and Culture project, she asked a woman from southern Brazil about the reason some people in her culture viewed owls the way they did. The woman responded, "Elas encantam o ambiente." *Because they enchant the environment.*

Owls are not omnipresent for us in the way songbirds are, those beautiful warbling flowers. But they're present for us in some deeper way or place, where night lives inside us, and also out there in the dark, embodying the unknown and the unknowable.

By provoking powerful reactions in us, owls offer a cultural service. "They kind of stir the pot for people so that we have a full range of experience and feelings, get spooked, feel awe and wonder," says Felice Wyndham. "That's a service, too." It's also a reminder of how vital it is for conservationists to consult with Indigenous leaders and to try to integrate what is culturally important to people about owls, she says. "Conservationists rooted only in natural science training and focusing only on ecosystem services may forever find themselves in a shallow paradigm that doesn't quite work on the ground until they engage with and respect local community perspectives and relationships with owls."

One evening in August, around sunset, I watched the release of a Barred Owl that had been cared for at the Wildlife Center of Virginia. The release took place on a mountaintop outside of Charlottesville, 600 acres of prime owl habitat, mixed forest crawling with small mammals. The young Barred Owl had been brought to the center a few months earlier, in late April. The official story was that she had fallen from a nest, but in fact landscapers had cut down the tree, and the baby owlet had tumbled to the ground. She was only a couple of weeks old, and rescuers from the center tried to renest her, setting her in a wicker Easter basket strapped to a tree. But there was no sign of Mom or Dad, and the next day she was again on the ground. So the staff brought her to the center and placed her with five other baby Barred Owls and three adults.

Now she was ready to go back to the wild. The center's veterinarian, Karra Pierce, took her to Piney Mountain in a dog crate. The center tries to release adults in the place they were found. "That way, they already know where to hunt, sleep, get water," says Pierce. "This one we'll release in good habitat and hope for the best." She's in top shape and well fed, but it will take time for her to find fruitful hunting grounds and safe spots for roosting and sleeping.

Pierce puts the crate on the ground in an open spot. She's a little nervous about this release. "Others in the past have not gone as planned," she says. "An eagle we released to great fanfare, with press waiting and a planned aerial salute, just scuttled into some scrubby undergrowth nearby and wouldn't fly."

A small group of us gather in a semicircle some distance from the crate. The view from the mountain is spectacular, the city below us, the forest spanning out from the mountaintop, the setting sun. One person asks if we should chant or sing a song or something. No, says Pierce. In fact, it's best to be as quiet as we can. The quieter we are, the more likely the owl will do what she wants to do and what Pierce wants her to do, to fly off and return to her life in the wild. So we stand in stillness as best we can to make ourselves invisible.

Pierce opens the crate, and the owl launches straight out, flapping hard into the woods. She settles in a high tree some distance from where we're gathered, but still in eyeshot. "She's so smart," says Pierce. "She's surveying the land. She loves this kind of dense forest. She

Barry the Barred Owl

might find this a good habitat, a good home, but if not, she will find one elsewhere. With any luck," Pierce says, "I'll never see this bird again."

That same evening, hundreds of people gathered in Central Park to hold a vigil for Barry, the beloved Barred Owl that had captivated New Yorkers before her death. People brought flowers and wrote messages in chalk beneath her hemlock tree in the Central Park Ramble, where she roosted. They stood quietly to attend her absence but felt her presence everywhere. An artist drew a portrait of her on the sidewalk with the quote "Nothing in a forest ever dies . . . It just changes shape."

There or not there, Barry would enchant the landscape. And, like all owls, remind us that we are always perched on the edge of mystery.

Afterword

Saving Owls

Protecting What We Love

Birds have been on this planet in one form or another for more than a hundred million years, diversifying over time into a breathtaking array of species, from shovelers to frogmouths, penguins to potoos, hummingbirds to hornbills, umbrellabirds to Emus. But today almost half the world's species of birds is declining.

What about owls? Which species are thriving? Which are in trouble and why? What can we do to save them?

The wide-ranging generalists of the owl world, the Barred Owls, Ural Owls, and Eastern Screech Owls, seem to be doing well, expanding their populations. These birds with flexible diets and habits will try almost anything to make a living, and can thrive even in dense and chaotic urban environments. Their adaptability and ingenuity at making use of the human-shaped world might be viewed as a positive sign.

Others, such as Little Owls, Northern Saw-whet Owls, and Great Horned Owls, are holding their own in the face of change. But many owl species are at risk from the disappearance of the big old-growth trees that once harbored their nesting hollows and the vast meadows and grasslands that served as their hunting grounds, from the threat of invasive species and rodenticides, and from the widespread effects of climate change. The losers in the drama of evolution and extinction will likely be the specialists limited to narrow ecological niches and small geographic ranges, island species, forest-dependent species, and those that have lost most of their habitats.

Owls living on islands are perhaps the most vulnerable in the world. Most of the owl species tottering on the threshold of extinction today are island dwellers, among them the Siau Scops Owl, a tiny brown owl with ear tufts, confined to the vestiges of forest on the little island of Siau in Indonesia, and the endangered Tasmanian Masked Owl, which nests in big hollows in the island's old-growth eucalypts—a rapidly diminishing resource.

The loss of native habitat is the single greatest threat to owl populations. In the past centuries, urbanization, deforestation, and agricultural development have stripped many owls of their forest and grassland habitats. What remains is often degraded by shifts in plant species from native to invasive, which in turn means that the small mammals that feed on native grass seeds decline, and along with them, the owls that prey on them.

Human activity around the world is carving nearly all owl habitats into isolated, island-like fragments, from the moist tropical woodlands of Brazil's Atlantic Forest to the grasslands of Montana, putting at risk local populations, such as the Forest Owlet of India, the Chestnut-backed Owlet of Sri Lanka, and the Long-whiskered Owlet of the Peruvian Andes. The two new owl species discovered in Brazil, the Alagoas Screech Owl and the Xingu Screech Owl, are both at risk of extinction. Despite massive conservation efforts, populations of the threatened Northern Spotted Owl are now at an all-time low. The US Forest Service just categorized the Ferruginous Pygmy Owl as threatened. The

Snowy Owl

Burrowing Owl has been listed as a species of national concern in the United States, with pronounced declines in the Great Plains, and is endangered in Canada. Even Long-eared Owls, considered of "Least Concern" on the official IUCN Red List of Threatened Species, are by some accounts tumbling toward calamity in some places.

Worries run deep over the ethereally beautiful Snowy Owl. A decade ago, the global population of Snowies was estimated at 200,000, including separate populations nesting in Russia and in the Northwest Territories of North America. But through satellite tracking and genetics, scientists came to realize that the number was probably much lower. The Snowy Owl populations, it turns out, are not separate but rather a single population moving around across the entire Arctic. The same pair, for instance, might nest in Russia one year, in the Northwest Territories the next year, and on Baffin Island the following year.

Now the population estimate for the Snowy Owl is around only 30,000 adults, and the bird is classified as vulnerable.

The idea that Snowy Owls, creatures lodged so deeply in our psyches, might vanish—to have that magic and then to lose it—is unfathomable and points to the urgency of doing everything we can to save these birds.

As I write this, Denver Holt is in the Arctic searching for Snowy Owl nests near Utqiaġvik, Alaska. This area northwest of Prudhoe Bay, between the Chukchi and the Beaufort Seas, has for decades been considered the most reliable breeding site for Snowies in the United States. Holt has been coming here every June for more than thirty years to try to grasp what's going on with Snowy Owl breeding trends and to monitor the lemming populations. Some summers he has found up to fifty nests. In 2021 and 2022, he found only one nest. Lately, nearly all the slight rises in the tundra that the owls have traditionally used as nesting sites have been vacant.

Looking for Snowy Owl nests is hard work. The Arctic is a place of unremitting challenge, even in summer when its wide-open landscapes are light-washed all day, with a subtle beauty you must lean in to see, bright spots of ground-hugging flowers speckling the tundra, yellow Snow Buttercup or pink Woolly Lousewort. Farley Mowat wrote about the pull of the vast Arctic landscape as a sort of disease, "a fever [that] has no effect on the body but lives only in the mind, filling its victim with a consuming urge to wander again . . . through those mighty spaces." But for Holt, the wandering in search of nests is exhausting, long slogs through sodden permafrost that sucks at your boots.

It's hard to know why the owls aren't breeding in Utqiaġvik in the numbers they once were, he says. "The trend is downward for Snowy Owls and slightly downward for lemmings, too." The lemming numbers still fluctuate, but the highs are not as high as they used to be, and it has been one bad year after another. The "normal" oscillations are being affected by a bigger change. The ice is thinning, the permafrost is melting, the snow is changing, all of which affects the lemmings. "If the lemming populations are in trouble, the implications for the food web are radical," he says. "The little mammals—and the big owls they feed—are a barometer of the pressure from climate change and environmental calamity. But we just don't know the details." He can't help but worry about the confusion over the numbers and the possibility of deep and threatening change.

Scientists have just begun to glean the impacts of human-induced climate change on owls. Shifting temperature can uncouple the availability of prey with the timing of breeding and nestling needs. Migratory owls may be particularly vulnerable, as their migrations are timed to coincide with the emergence of insects and other events in ecosystems that are affected by temperature. Warming temperatures can hobble the larders, or "freezers," that some birds, like Eurasian Pygmy Owls, use to hoard perishable food for overwintering. Prolonged drought can cause risky delays in nesting for some species, such as Burrowing Owls. A shifting climate can radically alter habitat and shift the ranges of owls and of their prey. Species dependent on high-elevation forests, like Flammulated Owls, might see their ranges contract. Scientists predict that within this century, aspen forests may all but disappear in many parts of North America, and along with them, the cavities so many owls depend on for nesting.

"The owl of Minerva spreads its wings only with the coming of dusk." According to some interpretations, Hegel's famous (and perhaps only) mention of owls suggests that human understanding comes only when it's too late.

It's not too late, says David Johnson. But with so many owl species declining so rapidly, there's no time to waste. "The clock is ticking on owl extinctions," he says. Owls are disappearing every day, and the changes in forests, grasslands, and tundra are almost certainly happening too fast for owls to adapt.

What should we do?

Everything in our power.

People all around the world are making efforts to mitigate the disasters—to slow deforestation and development, to eliminate or reduce the use of rodenticides and other poisons, to eradicate invasive species, to build nest boxes in places where the trees that once offered nesting cavities and hollows for owls have vanished. Researchers and citizen scientists are making the boxes from beautiful native wood in Japan, from soybean oil barrels in Russia, from pine ceiling boards in South Africa, and from 3D printers in Australia.

One very important step is getting a handle on the populations of owls, says Holt. "The truth is, for most species, we don't know the numbers." Owl populations are not easy to monitor. Some species move around with the seasons and from year to year in pursuit of pockets of reliable prey, and their movements are still not well understood, so they're hard to keep track of. Most studies collect data for a few seasons at most, "and there's no way that short-term studies can provide robust data," says Holt. "There are just too many fluctuations. To get a true picture of populations, you need to monitor them for a lifetime."

Not many researchers do this. Some of the world's longest-running studies have been conducted by Pertti Saurola, who has monitored owl populations in Finland for well over a half century. Humans are changing the environment so rapidly that annual systematic population monitoring is essential, says Saurola. In Finland, he has organized a small army of highly skilled volunteer banders to do the job. For most owl species, including Eurasian Eagle Owls, Great Grays, and Boreal Owls, the trends have been negative. The good news? Ural Owl and

Tawny Owl populations have held steady or are slowly increasing thanks to the construction of nest boxes by volunteer banders to compensate for the loss of old trees with big cavities.

The methods used for counting birds in many places, like breeding bird surveys and Christmas bird counts—a census of birds performed every December by volunteer birdwatchers—aren't tailored to detect owls. They're aimed at counting daytime singing birds. "We need to standardize methods of monitoring and gear them specifically to owls," says Holt. "That's the only way we'll get the reliable information we need on population status, habitat, and long-term trends to create sound conservation programs."

Citizens can help with this effort. Since 1991, Jim Duncan and his wife and fellow zoologist, Patricia Duncan, have led volunteer nocturnal owl surveys in Manitoba, Canada, during the spring breeding season. They launched the program because they wanted not only to fill the gap in existing monitoring programs that don't have suitable owl survey methods, but also to give citizens personal experience with owls. The program started small, but then word spread organically, and it grew and grew, says Duncan. "Owls as a group of birds just seem to have a disproportionate resonance or impact on people." Over a 25-year span, the program drew 900 volunteers, who detected more than 6,300 owls of 11 different species. The Duncans coordinated everything out of their farmhouse kitchen in Manitoba. The volunteers came from all walks of life. "The top performers, the people who came back to survey year after year, included a nuclear chemist, a poet, a park biologist, two women farmers, a young student, and some retired folks," says Duncan. "It was a very eclectic group." Their dedication impressed him. One volunteer ran into trouble while conducting a survey on route number 13, part of Duncan's study area for nesting Great Grays.

"His report said, 'I got to stop seven, and I could see lights from a van up ahead, and then the lights went off. While I was surveying for owls, I heard a gunshot, and a bullet whizzed by my head. So I'm really sorry, but I skipped the next stop and continued on with stop nine. I'm so sorry I missed that spot.'

"And I said to him, 'Oh, man, why did you stay there? It was probably poachers defending their poaching territories. You should have just gotten the heck out!' The dedication and the craziness this guy had for the survey—he just kept doing it!"

The data the volunteers collected on the distribution and abundance of owls in different habitats and their population trends have been used widely in research papers and conservation efforts. "It's a way of monitoring owls over big geographic areas over multiple decades," says Duncan. "And it connects people to nature and to owls, so they're more likely to support and demand conservation actions to preserve habitat. In Canada, these so-called owl prowls have proliferated in all corners of the country. Maybe one day they'll catch on elsewhere."

To save owls, we need to broaden and deepen our understanding of their populations and the habitats they live in, and that means studying the birds in all the ways described in this book. It means tracking them long term and researching their demographics, their behavior, biology, and ecology, season after season, year after year. It means safeguarding the places they nest and roost. And perhaps most important, it means educating people about the birds' nature and their need for protection.

"Humans are pretty much the biggest problem for owls around the world," says Karla Bloem. The drive to develop land for agriculture, housing, and industry is destroying what few refuges remain for owls.

And in some parts of the world, owls are still actively persecuted or hunted or captured and killed for their symbolism.

In some cultures, owls are literally loved to death. In India, on the night of Diwali, the Festival of Lights, Lakshmi, the goddess of wealth and prosperity, is said to travel the earth, visiting homes lit with lamps. In Hindu mythology, owls are considered the vahana (or vehicle) of Lakshmi. Some people believe that killing an owl—although it's illegal to do so—will trap Lakshmi in their house, bringing them good luck and wealth all year round. As a consequence, the Festival of Lights brings death to thousands of owls.

Efforts to shift cultural attitudes can go a long way toward protecting species in these places.

Raju Acharya has worked for decades to change attitudes in his native Nepal. In his country, he says, owls are believed dumb and unlucky. They're hunted for food and for enjoyment. "The government provides much less protection for owls than for snow leopards, rhinoceroses, and other threatened species." So Acharya brings festivals that feature dances, games, and other cultural activities to rural areas of his county to draw people in, and then pairs these activities with education about the twenty-three species of owls that live in Nepal. He has conducted hundreds of conservation camps in schools, especially near the borders between China and Nepal, held workshops for political leaders, and published conservation materials and owl newsletters, reaching millions of people.

In parts of the world where owls are shot on sight because they're considered evil omens or harvested for traditional medicines or cures, conservationists are focusing on education, especially the education of children.

In the 1990s, the Zambian Ornithological Society decided to survey children about their views of owls in a country where the birds are often killed by farmers and villagers. The society asked the children to write down the stories they had heard about owls, to draw pictures, to take questions home to ask their elders, and to write about their answers. The resulting flood of entries was so compelling that the society made them into a little book, *Owls Want Loving.* The book is threaded through with traditional knowledge and natural history information about the role owls play in providing valuable ecological and economic services. "It's one of the best educational tools I've seen," says David Johnson, "and it's now used in primary schools throughout Zambia."

In South Africa, a project launched by Jonathan Haw as a way of dealing with a surging rat population in Johannesburg has also worked as an effective owl educational program. Owlproject.org started twenty years ago with the purpose of raising nest boxes in townships to address the rat problem and, at the same time, to change attitudes about owls. The project team worked with schools to introduce Black South African schoolchildren to the natural wonders of owls and to their potential as partners in controlling rats in the impoverished communities where the children live.

"People are intimidated by things they don't know," says Kefiloe Motaung, a manager for the project. "The best way to tackle that is by educating people on the value of owls." Since the project began, some 260,000 schoolchildren have been a part of it, dissecting owl pellets, listening to educational owl talks, seeing baby owlets hatched in the nest boxes, helping to rehabilitate and release owls, and learning about the role owls play in taking care of the rat problem. "Once they finish the program, they get a certificate and they become our ambassadors," says Haw. "The beauty of this is that these kids go home and educate their parents, so it's a sort of bottom-up education program. The parents then often come in and do an owl pellet dissection course themselves."

The project follows up with questionnaires for both children and

Children in an orphanage in South Africa, with barn owls brought by Jonathan Haw and owlproject.org

their parents. Do you think owls are scary? Do you think they can be used for medicine? Do you think they're good for nature? "We had about two thousand children answer these questionnaires," says Haw. "We compared children who had participated in this project a year before—so they've had a chance to absorb or digest it—and kids who hadn't participated at all. The results were fascinating. The kids who participated in the program said owls couldn't be used for medicine. They weren't scary. They were good for the environment. And the kids who hadn't participated, well . . . the answers weren't the same. The discrepancy was huge."

The program also collaborated with an American organization called RATS (Raptors Are the Solution), which works to reduce the use of rodenticides, creating a beautiful, colorful poster with pro-owl messages in Zulu and ten other languages in South Africa. "We've put it up in townships all over the place for kids to read and have a look and see that this is about the message," says Haw. The project also enlisted more than 200 South African taxi drivers to put bumper stickers on their cabs with this simple message: Owls Eat Rats.

In places where owls are viewed in a negative light, conservationists are also working to heighten awareness of their economic value. One of Milan Ružić's favorite arguments to Serbian farmers for keeping Long-eared Owls in their fields goes something like this: Rodents eat thousands of kilos of grains in farm fields every year. An owl family with three chicks preys on some 8,000 rodents a year. With one breeding pair of Long-eared Owls on the property, a farmer can save 16,000 kilos of grain, which earns about €700 a month.

What jury would convict an owl after that piece of testimony?

In Italy, where Little Owls are still often linked with death and feared and harassed, naturalist and writer Marco Mastrorilli has created owl-friendly trails, owl workshops, and a massive owl festival. Mastrorilli has written dozens of books about owls of all kinds, many for children. In 2016, he built two Sentieri dei Gufi, "Trails of Owls," in northwestern Italy. The low path runs through chestnut groves and around the village of Venaus, "kingdom of the Tawny Owl," and is populated with striking sculptures of owls created in a competition. The other, a high-elevation trail that winds through larch woods in the foothills of the Alps near the border with France, is excellent Boreal Owl habitat. Both paths are studded with informational placards and signs about owls. On the upper trail, we encountered a group of thirty children rambling from signpost to signpost. When Mastrorilli introduced himself, they stopped and listened, entranced by his descriptions of the Boreal Owls and the Black Woodpeckers that drill the nesting holes used by the owls.

Mastrorilli's mission to educate people about owls extends to people with visual disabilities. At a 2009 meeting of the Unione Italiana dei Ciechi (the Italian Union of the Blind and Visually Impaired) in

Milan, he conducted a workshop for twenty people with visual impairments and their partners, explaining the natural history of owls and then offering opportunities to explore through touching feathers, pellets, feeders, and nest boxes. That night, he took the participants to the wooded Indro Montanelli Garden in the center of Milan to listen for Tawny Owls. "It was very emotional for me," he says. "After so many years visiting owls at night, to experience these birds with people for whom night is a constant and who had keener listening skills than I have . . . well, it was such a different reality. These people had no difficulty understanding how owls fly well in the night." Afterward, he created the Notte da Ascoltare (Night to Hear), taking people with visual impairments into the field to listen for wild owls and learn about them.

But Mastrorilli's biggest effort to foster a love of owls among the general public was creating the Festival dei Gufi with his partner, Stefania Montanino. The festival grew in attendance from just a few hundred people at its launch to 25,000 at its peak in 2015. That year, the queue to enter was more than a mile long. The Festival dei Gufi is the only owl festival of its size in the world, and it's aimed at educating people, focusing on the natural history of owls and why they should be revered, not feared. Rehabilitation specialists bring ambassador owls to show people their remarkable adaptations from a safe distance. Workshops teach the ecological and economic benefits of owls. Most of all, the festival is a celebration of all things owl, with thematic food; an international contest for owl wood carvings and owl body painting (with truly spectacular results); owl wine, beer, and cakes; a Snowy Owl ballet; even porta-potties painted with owls, along with handmade owl crafts from more than a hundred artists, including mugs inscribed with the slogan "I gufi non portano sfiga" (Owls don't bring bad luck).

Sometimes owls are so deeply loved in a place that people will go to truly extraordinary lengths to keep them around.

I think of the lifelong efforts of Sumio Yamamoto to keep the beloved Blakiston's Fish Owl alive in his native Japan. I met Yamamoto twenty years ago on the island of Hokkaido. At that time, he had already been studying the fish owl and working to conserve it for three decades. He explained to me that the bird was once widespread, but the cutting of old-growth trees for logging and development and the construction of dams and river channels, which destroyed native fish runs, had hit the birds hard, reducing the population to fewer than a hundred individuals in 1970. Since then, Yamamoto has overseen the building of hundreds of nest boxes to replace lost nesting sites, the planting of trees along riverbanks to create corridors of forests, and the construction of supplemental fishponds to improve available habitat. He has pioneered techniques for captive rearing of the endangered owl and fostered abandoned owl chicks. Together these measures have more than doubled the population, rescuing the owls from extinction in Japan.

For the past dozen years or so, Yamamoto has had a kindred spirit in Jonathan Slaght, who has partnered with his Russian colleagues Sergei Surmach and Sergei Avdeyuk to keep the Blakiston's Fish Owl population afloat in Russia. The researchers have worked to influence the actions of a logging company that builds access roads into the old-growth habitat favored by the endangered owl in the Primorye region of eastern Russia. The chief threat to owl habitat there is human use of these areas. People go in and fish, disturbing the owls or even killing them. The chief solution is reducing access. So Slaght and Surmach created a map of the region using GPS data to identify what parts of the landscape were most important for the fish owls. They gave the map of the fish owl "hot spots" to the logging company. "So if they're putting a new logging road in a river valley, they can refer to the map," says Slaght. "They can see, okay, this side of the valley has a patch of good fish owl habitat. We'll just put our road over here. And there's enough space in these valleys that this has been working." For roads that are already built, Slaght and Surmach have persuaded the company to reduce access to key stretches of road by putting up a huge berm of dirt blocking the path or by knocking out a bridge.

On tiny Norfolk Island, 800 miles from the east coast of Australia, researchers and citizens have made heroic efforts to save the Norfolk Island Morepork, a gorgeous little chocolaty-brown owl, about the size of a hand. "This is one of the most remarkable stories of conservation of a species because it got down to one bird," says Rohan Clarke, head of the Ornithology and Conservation Management Research Group at Monash University in Australia. "We knew we had to save it."

The Moreporks are still extremely rare, numbering at the most about thirty-five birds. The story of their dramatic rescue is ongoing and involves the efforts of a whole passel of people—researchers, park rangers, community members—pitching in to save the owls they love.

Norfolk Island is a place of sparkling blue waters, jagged cliffs, and gorgeous rainforests. But deforestation of the island over centuries and the removal of big, old trees that offered the Moreporks roosting and nesting hollows devastated the population of the little owls. When a team of ornithologists went to the island in the 1980s to look for Moreporks, they were able to find only a single female bird, named Miamiti by the islanders. The team met with the community to win their support, then brought to the island two male New Zealand Moreporks, the closest relatives to the Norfolk Island Morepork they could find. One male mysteriously disappeared, but Miamiti mated with the other, named Tintola. These birds went on to have their own offspring, and though the recovery was slow—the owls breed only once a year,

Norfolk Island Morepork

with a two-egg clutch—by 1996, when Miamiti died, she and Tintola had generated some fifty owls, each of them enchanting the landscape as their grandmothers or great-grandmothers had done, all the way back to Miamiti.

The rescue worked, but not for long. With only two founders, the population was highly inbred. "The entire owl population on Norfolk Island to the best of our knowledge is derived from that single pair," says Clarke, with lots of "backcrosses," Tintola mating with subsequent daughters, and other closely related birds mating with each other. "With a population that small, there's only so many opportunities, so the level of inbreeding you get has to be close to the maximum."

Enter Flossy Sperring, a graduate student advised by Clarke. Her charge at Norfolk Island was to understand as quickly as possible the genetics, ecology, and breeding behavior of the owls, as well as their interactions with invasive and native species, what the problems were, and how to address them. Now Sperring knows every owl living on Norfolk Island, knows its age, sex, lineage, its mate (if it has one—there are just seven pairs), where it roosts, nests, and hunts, its diet, and its name (Owlbert, Owlfonzo, Owlfreda, Owlex, etc.). In Sperring's first year of work, a small miracle occurred: two owl chicks fledged. It was the first successful breeding attempt on the island in almost a decade. Unfortunately, the birds no longer seem to be hatching young, so another rescue plan is in place, to bring in more New Zealand Moreporks.

Why put all this energy into a struggling population of little owls on a remote island? For the same reasons one might strive to protect owls anywhere in the world. To preserve biodiversity and the ecological fabric of the island, and because the owl is beloved by the island's human community, so much so that people spend hundreds of hours volunteering to monitor them and to help make decisions about their future. "Every species deserves protection," says Sperring. Every species matters to our concepts of beauty, ecological integrity, the sacredness of life. "And," she adds, "this one is so charismatic and cute."

Even as we are all part of the problem, we are also part of the solution.

What can an individual do?

I put the question to a range of owl experts: Do what you can, they said. Advocate for the preservation of critical habitat and against forces that destroy it—logging, urbanization, agricultural development. Create owl habitat yourself, safe places for owls to roost and nest. Check live trees for hollows before felling or trimming them, and leave snags and other dead trees standing when they don't pose a risk. Put up nest boxes. Use traps to control rats and mice rather than poisons. Learn what you can about owls around you. Who lives there? And where do they live? What are the threats facing them? Go out to try to find your local owls, but be aware of the impact you're having. Using playback to draw in owls can stress already stressed birds. Be courteous and respectful. Volunteer at raptor rehabilitation programs, at banding stations, and in the field with owl researchers. In 2019, an international group of scientists evaluating raptor research and conservation efforts around the globe determined that owls were the least-studied group of raptors and therefore the highest priority for future research. Dave Oleyar, who was part of that group, encourages people to join the effort. "Working on an owl project will forever change the way you think about trees, forests, and the amazing things happening at night," he says. Jim Duncan agrees. Get to know owls in whatever way you can, he says. "Exploring the lives of these creatures is always a rich and rewarding activity and can be full of surprises. To young people interested in a career, I want to say that owls are way more complex than we think, way cooler than we can imagine. And there's still so much to explore, to discover. It's an exciting place to be."

Last summer, conservation biologist and photographer Day Scott did what she felt she could do for owls. She joined a group of fellow biologists and citizen scientists trying to understand the long-term effects of climate change on small forest owls in the Chiricahua Mountains of southeastern Arizona.

These mountains belong to the "Sky Islands," isolated mountain ranges that are breathtakingly beautiful and one of the planet's hot spots for owl richness. Four different geographical regions meet here, creating a biodiversity hot spot not just for owls but for all kinds of wildlife. "It's a mecca for animals," says Scott, and for the people studying them. "We were the owl people, but the ant people were there from Germany, the hummingbird people were there, the bat people, the lizard people."

The owl project Scott joined is the brainchild of Dave Oleyar and a collaborative effort between HawkWatch International and the Earthwatch Institute, which draws volunteers of all ages and from all walks of life to assist in the research. The goal of the project is to better understand the region's six species of small, secretive owls, how they're affected by climate change, forest type, and forest management, and how to keep their populations afloat. Some of these little owls are particularly susceptible to climatic shifts: Flammulated Owls, which are migratory and feed primarily on insects, and Whiskered Screech Owls, which live at the northern edge of their range in these mountains. But the others may also be at risk—Elf Owls, Northern Saw-whet Owls, Northern Pygmy Owls, and Western Screech Owls. The project also focuses on tree hollows—how many there are in different types of forest, how long they last, and how these factors affect the owls.

Under the leadership of Oleyar, Scott and several other biologists and volunteers surveyed for owls and their natural nesting cavities during the day. At night they trapped the owls in mist nets to band them. Navigating the terrain—a mix of canyons and high-elevation coniferous forests, rocky, uneven, steep in places, with thick vegetation and creeks to cross—was difficult in the daylight. At night, in the deep darkness, it could be treacherous.

Two years ago, while Scott was working as an environmental educator in Wyoming, she was injured in a vehicle crash with a herd of Pronghorn. The accident left her with a traumatic brain injury. Because her disabilities limit her daily activities, she knew that joining the field team would be challenging. "But also joyful," she says, and collecting the data would help her own research on forest owls and contribute to the larger climate change project. Scott also considered her influence as a Black woman with a disability, and the importance of marginalized groups being represented in the conservation science community. "Humans play such a huge role in conservation," she says. "We have to work together to save every living thing."

For six weeks that summer, Scott got up every morning at seven and joined the community science team traversing the mountainous landscape all day to search for owls and their nesting cavities. Team members noted the characteristics of the tree hollows—their height, the direction they faced, whether they were natural or excavated, the tree species and whether the tree was alive or dead. They collected data on the vegetation in the surrounding habitat, the diameter of trees, their density and canopy cover. For her own independent research project, Scott explored the relationship between elevation and the diversity and number of owls heard during nocturnal surveys. Every night from eight to midnight, she participated in raising mist nets to trap the owls and then helped to process them, taking measurements and photographs, shining a black light on their wings, and banding them before releasing them back into the wild.

One night, the team captured an Elf Owl, the world's smallest owl. Scott held the bird during the processing. "It was amazing," she says. "When you think of an owl, you think of a big bird, like a Great Horned Owl." This owl was no bigger than a sparrow and could fit easily in the palm of her hand. Elf Owls were the hardest to catch. "We could barely see them flying," she says. "It was pitch black, and they weigh less than forty grams. They would go over the net, under the net, around the net. They're very smart." The team managed to catch only four of the little owls during the whole season.

"Having this tiny living thing in my hand, I started thinking about everything I knew about owls," says Scott. "Their feathers, for instance. This bird was all fluff, way tinier even than she seemed. When you blew on her belly to see if she had a brood patch, you could see just how little there was of her. Imagine if we had all those feathers on us, how much bigger we would look! I thought about how those feathers help her fly silently. As opposed to Wilson's Snipe, which makes so much noise when it takes flight. And her feet. So tiny, but also killer feet. Tiny killer feet.

"All these thoughts were going through my head while I had this bird in my hand, and I wanted to know, What are *you* thinking? I felt like I partially knew. She was so tiny and fragile, and we are so big (but also fragile, at least emotionally)," she says with a laugh. "I was looking at her and feeling like we had some kind of connection, because I was trying to care for her in the best way I possibly could. And whatever the importance of life was, right then it came down to this one owl I was holding in my hand."

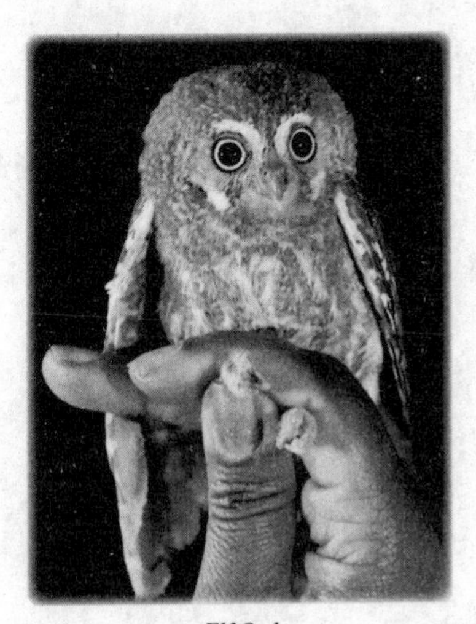

Elf Owl

Acknowledgments

In 2020, at the height of the pandemic, I decided to start work on a new book about owls. That difficult year when we were all marooned at home, I tuned in to a series of excellent online lectures offered by the International Owl Center in Houston, Minnesota, and curated by the center's director, Karla Bloem. Among the speakers was David Johnson, director of the Global Owl Project, who gave a talk on owls in myth and culture. After the talk, I emailed David to ask whether he might be willing to offer some guidance on the book. He responded immediately and, over the next two years, gave his time and unique expertise with extraordinary generosity. He invited me to observe his work on Burrowing Owls at the depot in Oregon and in Brazil and met up with me several times here in Virginia for interviews and consultation. He provided innumerable books and references and read over the entire manuscript at multiple stages, offering invaluable comments and suggestions. Throughout, David was an inspiration, passionate about owls, deeply knowledgeable, and endlessly giving, and I am very grateful to him.

In the same breath, I want to give huge thanks to Denver Holt, who took time from his busy schedule to show me his owl research in Montana and to share his vast knowledge and experience. As I found out, there's a very a good reason he's known as the "owl guy." Denver and his team at the Owl Research Institute, Beth Mendelsohn, Steve Hiro, Chloe Hernandez, Solai Le Fay, Lauren Smith, and Jon Barlow, took me into the field around Charlo and

Missoula for a full week and showed me what they do and how they do it, answering endless questions with great patience.

In Brazil, Priscilla Esclarski and Gabriela Mendes were knowledgeable guides to the Burrowing Owls of Maringá and offered helpful additions to the story. Many thanks to them and to team members Thaís Rafaelli and Vinícius Bonassoli.

I also want to thank Raphael Sobania for taking me out to see those neotropical screech owls and the family of Striped Owls in Curitiba, Brazil.

Karla Bloem, director of the International Owl Center, was a beacon on this project, referring me to owl experts all over the world, participating in numerous interviews, and offering helpful suggestions at every stage.

Karla, David Johnson, Denver Holt, and Dave Oleyar all gave the entire manuscript a thorough read and helped me avoid owl-related gaffes. Their astute comments and catches were tremendously helpful. Any lingering errors in the text are mine and mine alone.

Deep thanks to Julie Kacmarcik for facilitating my time at the Powhatan banding station, for many helpful conversations, and for that wonderful trip to see the barn owl family in Linville, Virginia. My gratitude, also, to Kim Cook and Diane Girgente for showing me the ropes of banding Northern Saw-whet Owls.

I'm extremely grateful to Amanda Nicholson and Karra Pierce for the time they gave me at the Wildlife Center of Virginia and for answering innumerable follow-up questions. A special tribute of thanks to the center's stellar surrogate father owl, Papa G'Ho, who sadly died in November 2022. I also appreciate Brie Hashem and Sarah Cooper of the Rockfish Wildlife Sanctuary, who provided a wonderful tour of the owls and other rescue animals at the sanctuary and discussed their training strategies.

Marco Mastrorilli and Stefania Montanino hosted me for three lovely days of owls and owl talk outside of Turin, Italy. They introduced me to the Tawny Owls that live in the woods near their home and showed me the two marvelous Sentieri dei Gufi, the "Trails of Owls," that Mastrorilli created to introduce people, especially children, to the wonders of owls. And as a side trip, they drove me into the French Alps, where—thanks to the sharp eyes of Stefania—we spotted Griffon Vultures. *Grazie di cuore a entrambi.*

I spoke and corresponded with a parliament of ornithologists, owl scientists, researchers, educators, naturalists, and rehabilitation specialists, who provided generous help on this project. Special thanks go to Lori Arent, Rob Bierregaard, Rohan Bilney, Mark Bonta, Reed Bowman, Gail Buhl, Joyce Caldwell, Nicholas Carter, Christopher Clark, Rohan Clarke, Jonathan Clarkson, Raylene Cooke, Sergio Cordoba Cordoba, Steve Debus, Jim Duncan, Laura Erickson, John Fitzpatrick, Robyn Fleming, Nicole Gill, Yoram Gutfreund, Jennifer Hartman, Jonathan Haw, Rob Heinsohn, Tomás Ibarra, Rod Kavanagh, Christine Köppl, Sean Larkan, Véronique Laroulandie, David Lindo, Pavel Linhart, Craig Morley, Beth Mott, José Carlos Motta-Junior, José Luis Peña, Niels Rattenborg, Bob Reilly, Nicole Richardson, Alexandre Roulin, Milan Ružić, Pertti Saurola, Marjon Savelsberg, Day Scott, Andrew Skeoch, Jonathan Slaght, Roar Solheim, Chris Soucy, Flossy Sperring, Rada Surmach, Jean-François Therrien, Carlos Mario Wagner-Wagner, Mark Walters, Scott Weidensaul, John White, Felice Wyndham, Connor Wood, and William Young. Many of these individuals referred me to other owl specialists, provided resources, and read portions of the manuscript, offering excellent suggestions. My profound gratitude to them all.

My dear friend Miriam Nelson read several chapters of the book and offered her usual sage advice.

A number of talented photographers, researchers, and artists contributed images to this book free of charge as an educational offering, among them Ambika Angela Bone, Nick Bradsworth, Lynn Bystrom, Jonathan Haw and owlproject.org, David H. Johnson, José Carlos Motta-Junior, Julie Kacmarcik, Nathalie Morales, Amanda Nicholson, Marjon Savelsberg, Flossy Sperring, and Čeda Vučković. Thank you all so much for providing your beautiful images. Robyn Fleming generously shared her expertise and helped me track down cultural images of owls in the public domain at the Met.

Other professional photographers and artists offered their work at a much reduced cost. I am extremely grateful to Nick Athanas, Nathan Clark, Dan Cox of Natural Exposures, Melissa Groo, Jeff Grotte, David Lei, Matt Poole, Day Scott, Jonathan Slaght, Roar Solheim, and Brad Wilson. A special shout-out to Dan Cox for agreeing to share his extraordinary image of Northern Pygmy Owls inside their nesting cavity, and to Terresa White for allowing me to show her stunning sculpture *Dependent Arising: Owl and Lemming.* Warm gratitude to Pete Myers, who helped me select photos for the book; to Donna Lucey, who shared her expertise on photography and layouts; and to my dear stepmother, Gail Gorham, for her generous support of the images.

To my editor at Penguin Press, Ann Godoff, I owe deep thanks for her belief in this book and for the way she tended to it at every stage with characteristic intelligence, insight, and grace. Thank you, Ann, for nudging me forward in your gentle way and for helping me bring owls to light. Many thanks to Casey Denis for shepherding the book from conception to completion with her usual superb competence, including expert guidance on the photographs and other artwork; to Darren Haggar for the gorgeous jacket design; and to Victoria Lopez for her able assistance throughout the production process.

Many thanks to Guilherme Weinhardt Minetto for his excellent support in the early stages of the book, for transcribing and editing the recordings of interviews and offering sound counsel.

To my agent of more than thirty years, Melanie Jackson, I owe more than I can say. Through the decades, Melanie has helped me shape ideas for my books and guided their evolution with truly extraordinary insight and skill. My gratitude is endless for all the excellence she has brought to our work together and for all the good she has done for my life and my career. I consider myself lucky indeed to call her my agent and my friend.

Deep thanks to my family for their love and support, especially to my stepmother, Gail, for her generosity and enthusiastic encouragement, and to my sisters, Kim, Sarah, and Nancy. For years, Nancy and her husband,

Steve, have been the wind in my sails in every way. I am grateful beyond words for their presence in my life.

Finally, my profound love and gratitude always to my daughters, Zoë and Nelle, who above all have taught me how to be in the world, shown me the way with their wisdom and their light. They may have been owlets once, but they've taken wing and are now in full, beautiful flight.

Further Reading

General

M. Cieślak, *Feathers of European Owls: Insights into Species Ecology and Identification* (Uppsala, Sweden: Oriolus, 2017).

J. Duncan, *Owls of the World* (Baltimore: Johns Hopkins University Press, 2016).

Ela Foundation, "Proceedings of the 6th World Owl Conference, Pune, India, 2019," in *Ela Journal of Forestry and Wildlife* 11, no. 1 (January–March 2022).

P. L. Enríquez, ed., *Neotropical Owls: Diversity and Conservation* (Cham, Switzerland: Springer, 2017).

International Owl Center, internationalowlcenter.org.

H. Mikkola, *Owls of the World: A Photographic Guide*, 2nd edition (Richmond Hill, Ontario: Firefly Books, 2013).

D. Morris, *Owl* (London: Reaktion Books, 2009).

R. W. Nero, *The Great Gray Owl: Phantom of the Northern Forest* (Washington, DC: Smithsonian Institution Press, 1980).

Owl Research Institute, *The Roost*, annual newsletter (Charlo, Montana), owlresearchinstitute.org/the-roost.

J. C. Slaght, *Owls of the Eastern Ice: A Quest to Find and Save the World's Largest Owl* (New York: Farrar, Straus and Giroux, 2020).

M. Unwin and D. Tipling, *The Enigma of the Owl: An Illustrated Natural History* (New Haven, CT: Yale University Press, 2017).

Chapter 1. Making Sense of Owls

J. N. Choiniere et al., "Evolution of vision and hearing modalities in theropod dinosaurs," *Science* 372, no. 6542 (May 2021): 610–13, doi.org/10.1126/science.abe7941.

A. Duhamel et al., "Cranial evolution in the extinct Rodrigues Island owl *Otus murivorus* (Strigidae), associated with unexpected ecological adaptations," *Scientific Reports* 10 (2020): 14019, doi.org/10.1038/s41598-020-69868-1.

M. Hanson et al., "The early origin of a birdlike inner ear and the evolution of dinosaurian movement and vocalization," *Science* 372, no. 6542 (May 2021): 601–9, doi.org/10.1126/science.abb4305.

D. F. Lane and F. Angulo, "The distribution, natural history, and status of the Long-whiskered Owlet (*Xenoglaux loweryi*)," *Wilson Journal of Ornithology* 130, no. 3 (September 2018): 650–57.

A. Louchart, "Integrating the fossil record in the study of insular body size evolution: Example of owls (Aves, Strigiformes)," in "Proceedings of the International Symposium 'Insular Vertebrate Evolution: the Palaeontological Approach,'" ed. J. A. Alcover and P. Bover, *Monografies de la Societat d'Història Natural de les Balears* 12 (2005): 155–74.

G. Mayr, "Strigiformes," in *Paleogene Fossil Birds* (Heidelberg, Germany: Springer, 2009), 163–68.

G. Mayr, P. D. Gingerich, and Thierry Smith, "Skeleton of a new owl from the early Eocene of North America (Aves, Strigiformes) with an accipitrid-like foot morphology," *Journal of Vertebrate Paleontology* 40, no. 2 (2020), doi.org/10.1080/02724634.2020.1769116.

G. Mayr and A. C. Kitchener, "Early Eocene fossil illuminates the ancestral (diurnal) ecomorphology of owls and documents a mosaic evolution of the strigiform body plan," *Ibis* (August 7, 2022), doi.org/10.1111/ibi.13125.

C. McGrath, "Highlight: Adaptations that rule the night," *Genome Biology and Evolution* 12, no. 10 (October 2020): 1909–10, doi.org/10.1093/gbe/evaa199.

A. M. Patel and D. W. Steadman, "The Pleistocene Burrowing Owl (*Athene cunicularia*) from the Bahamas," *Journal of Caribbean Ornithology* 33 (2020): 86–94.

M. Wink and H. Sauer-Gürth, "Molecular taxonomy and systematics of owls (Strigiformes)—an update," in "World Owl Conference 2017," ed. Inês Roque, special issue, *Airo* 29 (2021): 487–500.

Chapter 2. What It's Like to Be an Owl

M. L. Allen et al., "Scavenging by owls: A global review and new observations from Europe and North America," *Journal of Raptor Research* 53, no. 4 (December 2019): 410–18.

T. Bachmann, H. Wagner, and C. Tropea, "Inner vane fringes of barn owl feathers reconsidered: Morphometric data and functional aspects," *Journal of Anatomy* 221, no. 1 (July 2012): 1–8.

A. D. S. Bala, "Auditory spatial acuity approximates the resolving power of space-specific neurons," *PLoS ONE* 2, no. 8 (August 7, 2007): e675, doi.org/10.1371/journal.pone.0000675.

A. D. S. Bala and T. T. Takahashi, "Pupillary dilation response as an indicator of auditory discrimination in the barn owl," *Journal of Comparative Physiology A* 186, no. 5 (May 2000): 425–34, doi.org/10.1007/s003590050442.

A. Boonman et al., "The sounds of silence: Barn owl noise in landing and taking off," *Behavioural Processes* 157 (December 2018): 484–88.

D. F. Brunton and R. Pittaway, "Observations of the Great Gray Owl on winter range," *Canadian Field-Naturalist* 85, no. 1 (January–March 1971): 315–22.

C. E. Carr and J. Christensen-Dalsgaard, "Sound localization strategies in three predators," *Brain, Behavior and Evolution* 86, no. 1 (September 2015): 17–27, doi.org/10.1159/000435946.

C. E. Carr and J. L. Peña, "Cracking an improbable sensory map," *Journal of Experimental Biology* 219, no. 24 (December 2016): 3829–31, doi.org/10.1242/jeb.129635.

C. J. Clark, "Ways that animal wings produce sound," *Integrated and Comparative Biology* 61, no. 2 (August 2021): 696–709, doi.org/10.1093/icb/icab008.

C. J. Clark, J. Duncan, and R. Dougherty, "Great Gray Owls hunting voles under snow hover to defeat an acoustic mirage," *Proceedings of the Royal Society B* 289: 20221164, doi.org/10.1098/rspb.2022.1164.

C. J. Clark, K. Le Piane, and L. Liu, "Evolution and ecology of silent flight in owls and other flying vertebrates," *Integrative Organismal Biology* 2, no. 1 (January 20, 2020): obaa001, doi.org/10.1093/iob/obaa001.

C. J. Clark, K. Le Piane, and L. Liu, "Evolutionary and ecological correlates of quiet flight in nightbirds, hawks, falcons, and owls," *Integrative and Comparative Biology* 60, no. 5 (November 2020): 1123–34.

A. Devine and D. G. Smith, "Caching behavior in Northern Saw-whet Owls, *Aegolius acadicus*," *Canadian Field-Naturalist* 119, no. 4 (October–December 2005): 578–79.

L. Einoder and A. Richardson, "The digital tendon locking mechanism of owls: Variation in the structure and arrangement of the mechanism and functional implications," *Emu* 107, no. 3 (2007): 223–30.

P. Espíndola-Hernández et al., "Genomic evidence for sensorial adaptations to a nocturnal predatory lifestyle in owls," *Genome Biology and Evolution* 12, no. 10 (October 2020): 1895–908.

P. Espíndola-Hernández, J. C. Mueller, and B. Kempenaers, "Genomic signatures of the evolution of a diurnal lifestyle in OkkStrigiformes," *G3 Genes|Genomes|Genetics* 12, no. 8 (August 2022): jkac135, doi.org/10.1093/g3journal/jkac135.

C. Gutiérrez-Ibáñez et al., "Comparative study of visual pathways in owls (Aves: Strigiformes)," *Brain, Behavior and Evolution* 81, no. 1 (February 2013): 27–39, doi.org/10.1159/000343810.

W. M. Harmening and H. Wagner, "From optics to attention: Visual perception in barn owls," *Journal of Comparative Physiology A* 197 (July 7, 2011): 1031, doi.org/10.1007/s00359-011-0664-3.

Y. Hazan et al., "Visual-auditory integration for visual search: A behavioral study in Barn Owls," *Frontiers in Integrative Neuroscience* 9 (February 2015): 1–12, doi.org/10.3389/fnint.2015.00011.

J. Höglund et al., "Owls lack UV-sensitive cone opsin and red oil droplets, but see UV light at night: Retinal transcriptomes and ocular media transmittance," *Vision Research* 158 (May 2019): 109–19.

J. W. Jaworski and N. Peake, "Aeroacoustics of silent owl flight," *Annual Review of Fluid Mechanics* 52 (January 2020): 395–420.

E. I. Knudsen and M. Konishi, "A neural map of auditory space in the owl," *Science* 200, no. 4343 (May 1978): 795–97.

E. I. Knudsen, M. Konishi, and J. D. Pettigrew, "Receptive fields of auditory neurons in the owl," *Science* 198, no 4323 (December 1977): 1278–80.

K. Krishnam et al., "Turbulent wake-flow characteristics in the near wake of freely flying raptors: A comparative analysis between an owl and a hawk," *Integrative and Comparative Biology* 60, no. 5 (November 2020): 1109–22, doi.org/10.1093/icb/icaa106.

B. Krumm et al., "Barn owls have ageless ears," *Proceedings of the Royal Society B* 284, no. 1863 (September 20, 2017): 20171584, doi.org/10.1098/rspb.2017.1584.

B. Krumm et al., "The barn owls' minimum audible angle," *PLoS ONE* 14, no. 8 (August 23, 2019): e0220652, doi.org/10.1371/journal.pone.0220652.

K. Le Piane and C. J. Clark, "Evidence that the dorsal velvet of barn owl wing feathers decreases rubbing sounds during flapping flight," *Integrative and Comparative Biology* 60, no. 5 (November 2020): 1068–79, doi.org/10.1093/icb/icaa045.

K. Le Piane and C. J. Clark, "Quiet flight, the leading edge comb, and their ecological correlates in owls (Strigiformes)," *Biological Journal of the Linnean Society* 135, no. 1 (January 2022): 84–97.

G. R. Martin, "Sensory capacities and the nocturnal habit of owls (Strigiformes)," *Ibis* 128, no. 2 (April 1986): 266–77.

G. R. Martin, "What is binocular vision for? A birds' eye view," *Journal of Vision* 9, no. 11 (October 2009): 14, doi.org/10.1167/9.11.14.

I. A. W. McAllan and D. Larkins, "A feeding technique of the Powerful Owl *Ninox strenua*," *Australian Field Ornithology* 22, no. 1 (January 2005): 38–41.

M. Mo et al., "Observations of hunting attacks by the Powerful Owl *Ninox strenua* and an examination of search and attack techniques," *Australian Zoologist* 38, no. 1 (January 2016): 52–58.

J. Orlowski et al., "Visual pop-out in barn owls: Human-like behavior in the avian brain," *Journal of Vision* 15, no. 14 (October 2015): 4, doi.org/10.1167/15.14.4.

J. Orlowski, W. Harmening, and H. Wagner, "Night vision in barn owls: Visual acuity and contrast sensitivity under dark adaptation," *Journal of Vision* 12, no. 13 (December 2002): 4, doi.org/10.1167/12.13.4.

R. Payne, "Acoustic location of prey by barn owls (*Tyto alba*)," *Journal of Experimental Biology* 54, no. 3 (June 1971): 535–73.

J. L. Peña and M. Konishi, "Auditory spatial receptive fields created by multiplication," *Science* 292, no. 5515 (April 13, 2001): 249–52, doi.org/10.1126/science.1059201.

F. T. G. Rocha de Vasconcelos et al., "LWS visual pigment in owls: Spectral tuning inferred by genetics," *Vision Research* 165 (December 2019): 90–97, doi.org/10.1016/j.visres.2019.10.001.

I. Sazima, "Lightning predator: The Ferruginous Pygmy Owl snatches flower-visiting hummingbirds in southwestern Brazil," *Revista Brasileira de Ornitologia* 23 (December 30, 2015): 12–14, doi.org/10.1007/BF03544283.

R. Solheim, "Caching behaviour, prey choice and surplus killing by Pygmy Owls *Glaucidium passerinum* during winter, a functional response of a generalist predator," *Annales of Zoologici Fennici* 21, no. 3 (1984): 301–8.

T. T. Takahashi, "How the owl tracks its prey—II," *Journal of Experimental Biology* 213, no. 20 (October 2010): 3399–408.

J. R. Usherwood, E. L. Sparkes, and R. Weller, "Leap and strike kinetics of an acoustically 'hunting' barn owl (*Tyto alba*)," *Journal of Experimental Biology* 217, no. 17 (September 2014): 3002–5.

A. Van den Burg and K. Koenraads, "Owls in the realm of avian anatomy," abstract, *World Owl Conference 2017: Book of Abstracts*, 51, woc2017.uevora.pt/wp-content/uploads/2017/11/WOC2017-ABSTRACTS.pdf.

H. Wagner et al., "Features of owl wings that promote silent flight," *Interface Focus* 7, no. 1 (February 6, 2017): 20160078, doi.org/10.1098/rsfs.2016.0078.

H. Wagner, T. T. Takahashi, and M. Konishi, "Representation of interaural time difference in the central nucleus of the barn owl's inferior colliculus," *Journal of Neuroscience* 7, no. 10 (October 1987): 3105–16.

J. Wang et al., "Aeroacoustic characteristics of owl-inspired blade designs in a mixed flow fan: Effects of leading- and trailing-edge serrations," *Bioinspiration & Biomimetics* 16, no. 6 (September 17, 2021): 066003, doi.org/10.1088/1748-3190/ac1309.

Y. Wu et al., "Retinal transcriptome sequencing sheds light on the adaptation to nocturnal and diurnal lifestyles in raptors," *Scientific Reports* 6 (September 20, 2016): 33578, doi.org/10.1038/srep33578.

Y. Wu and D. H. Johnson, "Evolution in gene sequences responsible for hearing, sight, and digestion in 99 species of owls, raptors, and passerines," abstract, in "Proceedings of the 6th World Owl Conference, Pune, India, 2019," *Ela Journal of Forestry and Wildlife* 11, no. 1 (January–March 2022): 1215, elafoundation.org/ela/wp-content/uploads/2022/08/EJFW-11-1.pdf.

M. Zhao et al., "Optimal design of aeroacoustic airfoils with owl-inspired trailing-edge serrations," *Bioinspiration & Biomimetics* 16, no. 5 (September 2021): 056004, doi.org/10.1088/1748-3190/ac03bd.

C. Zhou et al., "Comparative genomics sheds light on the predatory lifestyle of accipitrids and owls," *Scientific Reports* 9 (February 19, 2019): 2249, doi.org/10.1038/s41598-019-38680-x.

Chapter 3. Owling

K. M. Dugger et al., "The effects of habitat, climate, and Barred Owls on long-term demography of Northern Spotted Owls," *The Condor* 118, no. 1 (February 2016): 57–116, doi.org/10.1650/condor-15-24.1.

A. Grimm-Seyfarth, W. Harms, and A. Berger, "Detection dogs in nature conservation: A database on their world-wide deployment with a review on breeds used and their performance compared to other methods," *Methods in Ecology and Evolution* 12, no. 4 (April 2021): 568–79, doi.org/10.1111/2041-210X.13560.

R. J. Gutiérrez et al., "The invasion of Barred Owls and its potential effect on the Spotted Owl: A conservation conundrum," *Biological Invasions* 9, no. 2 (March 2007): 181–96, doi.org/10.1007/s10530-006-9025-5.

J. Shonfield and E. M. Bayne, "Using bioacoustics to study vocal behaviour and habitat use of Barred Owls, Boreal Owls and Great Horned Owls," in "World Owl Conference 2017," ed. Inês Roque, special issue, *Airo* 29 (2021): 416–31.

R. Surmach and P. G. Mametiev, "Contemporary technologies in Blakiston's Fish Owl research," 2nd International Conference on Northeast Asia Biodiversity, Baishan, China, August 27–31, 2019.

S. K. Wasser et al., "Using detection dogs to conduct simultaneous surveys of Northern Spotted (*Strix occidentalis caurina*) and Barred Owls (*Strix varia*)," *PLoS ONE* 7, no. 8 (August 2012): e42892, doi.org/10.1371/journal.pone.0042892.

C. M. Wood et al., "Early detection of rapid Barred Owl population growth within the range of the California Spotted Owl advises the precautionary principle," *The Condor* 122, no. 1 (February 2020): duz058, doi.org/10.1093/condor/duz058.

C. M. Wood et al., "Illuminating the nocturnal habits of owls with emerging tagging technologies," *Wildlife Society Bulletin* 45, no. 1 (February 12, 2021): 138–43, doi.org/10.1002/wsb.1156.

C. M. Wood et al., "Using the ecological significance of animal vocalizations to improve inference in acoustic monitoring programs," *Conservation Biology* 35, no. 1 (February 24, 2021): 336–45, doi.org/10.1111/cobi.13516.

C. M. Wood, R. J. Gutiérrez, and M. Z. Peery, "Acoustic monitoring reveals a diverse forest owl community, illustrating its potential for basic and applied ecology," *Ecology* 100, no. 9 (September 2019): e02764, doi.org/10.1002/ecy.2764.

C. M. Wood, S. M. Schmidt, and M. Z. Peery, "Spatiotemporal patterns of the California Spotted Owl's territorial vocalizations," *Western Birds* 50, no. 4 (December 2019): 232–42, doi.org/10.21199/WB50.4.2.

Chapter 4. Who Gives a Hoot

M. M. Delgado and V. Penteriani, "Vocal behaviour and neighbour spatial arrangement during vocal displays in eagle owls (*Bubo bubo*)," *Journal of Zoology* 271, no. 1 (January 2007): 3–10, doi.org/10.1111/j.1469-7998.2006.00205.x.

L. S. Duchac et al., "Passive acoustic monitoring effectively detects Northern Spotted Owls and Barred Owls over a range of forest conditions," *The Condor* 122, no. 3 (August 2020): 1–22, doi.org/10.1093/condor/duaa017.

T. Grava et al., "Individual acoustic monitoring of the European Eagle Owl *Bubo bubo*," *Ibis* 150, no. 2 (April 2008): 279–87.

L. A. Hardouin, P. Tabel, and V. Bretagnolle, "Neighbour-stranger discrimination in the Little Owl, *Athene noctua*," *Animal Behaviour* 72, no. 1 (July 2006): 105–12, doi.org/10.1016/j.anbehav.2005.09.020.

K. A. Kinstler, "Great Horned Owl *Bubo virginianus* vocalizations and associated behaviours," in "Proceedings of the Fourth World Owl Conference, October 31–November 4, 2007, Groningen, the Netherlands," ed. D. H. Johnson, D. Van Nieuwenhuyse, and J. R. Duncan, *Ardea* 97, no. 4 (December 2009): 413–20.

P. Linhart et al., "Measuring individual identity information in animal signals: Overview and performance of available identity metrics," *Methods in Ecology and Evolution* 10, no. 9 (September 2019): 1558–70, doi.org/10.1111/2041-210X.13238.

P. Linhart and M. Šálek, "Acoustic methods for long-term monitoring of birds: Individuality and stability in territorial calls of the Little Owl (*Athene noctua*)," abstract, in "Proceedings of the 6th World Owl Conference, Pune, India, 2019," *Ela Journal of Forestry and Wildlife* 11, no. 1 (January–March 2022): 1231, elafoundation.org/ela/wp-content/uploads/2022/08/EJFW-11-1.pdf.

V. Penteriani et al., "Brightness variability in the white badge of the eagle owl *Bubo bubo*," *Journal of Avian Biology* 37, no. 1 (January 2006): 110–16, doi.org/10.1111/j.0908-8857.2006.03569.x.

V. Penteriani et al., "The importance of visual cues for nocturnal species: Eagle owls signal by badge brightness," *Behavioral Ecology* 18, no. 1 (January 2007): 143–47, doi.org/10.1093/beheco/arl060.

V. Penteriani et al., "Owl dusk chorus is related to the quality of individuals and nest-sites," *Ibis* 156, no. 4 (October 2014): 892–95, doi.org/10.1111/ibi.12178.

V. Penteriani and M. del Mar Delgado, "The dusk chorus from an owl perspective: Eagle owls vocalize when their white throat badge contrasts most," *PLoS ONE* 4, no. 4 (April 8, 2009): e4960, doi.org/10.1371/journal.pone.0004960.

D. Stowell et al., "Automatic acoustic identification of individuals in multiple species: Improving identification across recording conditions," *Journal of the Royal Society Interface* 16, no. 153 (April 2019): 2018094020180940, doi.org/10.1098/rsif.2018.0940.

S. Yamamoto, "Unusual behaviors by male Blakiston's Fish Owls: Rearing of unrelated offspring and simultaneous courtship feeding," *Journal of Raptor Research* 56, no. 2 (June 2022): 253–55, doi.org/10.3356/JRR-21-11.

Chapter 5. What It Takes to Make an Owl

I. O. Adejumo, "Strategies of Owl Reproduction," in *Owls*, ed. H. Mikkola (London: IntechOpen, 2019), doi.org/10.5772/intechopen.82425.

F. M. Baumgartner, "Courtship and nesting of the Great Horned Owls," *Wilson Bulletin* 50, no. 4 (December 1938): 274–85.

C. Deppe et al., "Effect of Northern Pygmy-Owl (*Glaucidium gnoma*) eyespots on avian mobbing," *The Auk* 120, no. 3 (July 2003): 765–71, doi.org/10.1093/auk/120.3.765.

C. I. de Zeeuw and C. B. Canto, "Interpreting thoughts during sleep," *Science* 377, no. 6609 (August 25, 2022): 919–20, doi.org/10.1126/science.add8592.

P. Ducouret et al., "Elder barn owl nestlings flexibly redistribute parental food according to siblings' need or in return for allopreening," *American Naturalist* 196, no. 2 (August 2020): 257–69, doi.org/10.1086/709106.

F. R. Gehlbach and R. S. Baldridge, "Live blind snakes (*Leptotyphlops dulcis*) in Eastern Screech Owl (*Otus asio*) nests: A novel commensalism," *Oecologia* 71, no. 4 (March 1987): 560–63 (1987), doi.org/10.1007/BF00379297.

J. Hollingsworth and R. J. Bilney, "A possible case of infanticide and cannibalism in the Powerful Owl *Ninox strenua*," *Australian Field Ornithology* 34 (2017): 129–30, dx.doi.org/10.20938/afo34129130.

D. W. Holt, "Why are Snowy Owls white and why have they evolved distinct sexual color dimorphism? A review of questions and hypotheses," *Journal of Raptor Research* 56, no. 4 (December 2022): 1–15.

D. W. Holt et al., "Characteristics of nest mounds used by Snowy Owls in Barrow, Alaska, with conservation and management implications," *Ardea* 97, no. 4 (December 2009): 555–61.

D. W. Holt and M. Larson, "Natural nest-site characteristics of two small forest owls with implications for conservation and management," abstract, in "Abstracts from the 2014 annual meeting of the Society for Northwestern Vertebrate Biology, in cooperation with the Washington chapter of the Wildlife Society, Northwest Partners in Amphibian and Reptile Conservation, Researchers Implementing Conservation Action, and the Global Owl Project, held at the Red Lion Hotel, Pasco, Washington, 3–7 February 2014," *Northwestern Naturalist* 95, no. 2 (Autumn 2014): 148–49, doi.org/10.1898/NWNAbstracts_95-2.1.

D. W. Holt and W. D. Norton, "Observations of nesting Northern Pygmy-Owls," *Raptor Research* 20, no. 1 (Spring 1986): 39–41.

M. D. Larson and D. W. Holt, "Rope dragging technique for locating Short-eared Owl nests," *North American Bird Bander* 43, no. 2–3 (April–September 2018): 62–64.

M. Mo and D. R. Waterhouse, "Development of independence in Powerful Owl *Ninox strenua* fledglings in suburban Sydney," *Australian Field Ornithology* 32, no. 3 (September 2015): 143–53.

C. O'Connell and G. Keppel, "Deep tree hollows: Important refuges from extreme temperatures," *Wildlife Biology* 22, no. 6 (November 2016): 305–10.

J. Olsen et al., "Behaviour and family association during the post-fledging period in Southern Boobooks *Ninox boobook*," *Corella* 44 (2020): 61–70.

M. F. Scriba et al., "Linking melanism to brain development: Expression of a melanism-related gene in barn owl feather follicles covaries with sleep ontogeny," *Frontiers in Zoology* 10 (July 26, 2013): 42, doi.org/10.1186/1742-9994-10-42.

A. Webster et al., "Diet, roosts and breeding of Powerful Owls *Ninox strenua* in a disturbed, urban environment: A case for cannibalism? Or a case of infanticide?," *Emu—Austral Ornithology* 99, no. 1 (1999): 80–83, doi.org/10.1071/MU99009D.

Chapter 6. To Stay or to Go?

S. R. Beckett and G. A. Proudfoot, "Large-scale movement and migration of Northern Saw-whet Owls in eastern North America," *Wilson Journal of Ornithology* 123, no. 3 (September 2011): 521–35.

S. R. Beckett and G. A. Proudfoot, "Sex-specific migration trends of Northern Saw-whet Owls in eastern North America," *Journal of Raptor Research* 46, no. 1 (March 2012): 98–108.

J. Bowman, D. Badzinski, and R. J. Brooks, "The numerical response of breeding Northern Saw-whet Owls *Aegolius acadicus* suggests nomadism," *Journal of Ornithology* 151, no. 2 (April 2010): 499–506.

M. Cheveau et al., "Owl winter irruptions as an indicator of small mammal population cycles in the boreal forest of eastern North America," *Oikos* 107, no. 1 (October 2004): 190–98.

T. Curk et al., "Winter irruptive Snowy Owls (*Bubo scandiacus*) in North America are not starving," *Canadian Journal of Zoology* 96, no. 6 (June 2018): 553–58.

F. I. Doyle et al., "Seasonal movements of female Snowy Owls breeding in the western North American Arctic," *Journal of Raptor Research* 51, no. 4 (December 2017): 428–38.

P. Ducouret et al., "Elder barn owl nestlings flexibly redistribute parental food according to siblings' need or in return for allopreening," *American Naturalist* 196, no. 2 (August 2020): 257–69, doi.org/10.1086/709106.

D. W. Holt et al., "Snowy Owl (*Bubo scandiacus*)," version 1.0, in *Birds of the World*, ed. S. M. Billerman (Ithaca, NY: Cornell Lab of Ornithology, 2020), doi.org/10.2173/bow.snoowl1.01.

I. E. Menyushina, "Snowy Owl (*Nyctea scandiaca*) reproduction in relation to lemming population cycles on Wrangel Island," in "Biology and Conservation of Owls of the Northern Hemisphere: 2nd International Symposium," ed. J. R. Duncan, D. H. Johnson, and T. H. Nicholls, *General Technical Report NC-190* (St. Paul, MN: US Department of Agriculture, Forest Service, North Central Forest Experiment Station, 1997), 572–82.

H. C. Mueller and D. D. Berger, "Observations on migrating saw-whet owls," *Bird-Banding* 38, no. 2 (April 1967): 120–25.

C. M. Neri and N. Mackentley, "Different audio-lures lead to different sex-biases in capture of Northern Saw-whet Owls (*Aegolius acadicus*)," *Journal of Raptor Research* 52, no. 2 (June 2018): 245–49, doi.org/10.3356/JRR-17-28.1.

Project Owlnet, projectowlnet.org.

M. L. Pruitt and K. G. Smith, "History of Northern Saw-whet Owls (*Aegolius acadicus*) in North America: Discovery to present day," in "World Owl Conference 2017," ed. Inês Roque, special issue, *Airo* 29 (2021): 326–48.

W. Saunders, "A migration disaster in Western Ontario," *The Auk* 24, no. 1 (1907): 108–110.

R. Solheim et al., "Snowy Owl (*Bubo scandiacus*) males select the highest vantage points around nests," in "World Owl Conference 2017," ed. Inês Roque, special issue, *Airo* 29 (2021): 451–59.

P. A. Taverner and B. H. Swales, "Notes on the migration of the saw-whet owl," *The Auk* 28, no. 3 (July 1911): 329–34.

J.-F. Therrien et al., "Irruptive movements and breeding dispersal of Snowy Owls: A specialized predator exploiting a pulsed resource," *Journal of Avian Biology* 45, no. 6 (November 2014): 536–44.

J.-F. Therrien et al., "The irruptive nature of Snowy Owls: An overview of some of the recent empirical evidence," in "World Owl Conference 2017," ed. Inês Roque, special issue, *Airo* 29 (2021): 527–34.

J.-F. Therrien, G. Gauthier, and J. Bêty, "An avian terrestrial predator of the Arctic relies on the marine ecosystem during winter," *Journal of Avian Biology* 42, no. 4 (July 2011): 363–69.

J.-F. Therrien, G. Gauthier, and J. Bêty, "Survival and reproduction of adult Snowy Owls tracked by satellite," *Journal of Wildlife Management* 76, no. 8 (November 2012): 1562–67.

J. Wall et al., "Twenty-five year population trends in Northern Saw-whet Owl (*Aegolius acadicus*) in eastern North America," *Wilson Journal of Ornithology* 132, no. 3 (September 2020): 739–45.

C. S. Weidensaul et al., "Use of ultraviolet light as an aid in age classification of owls," *Wilson Journal of Ornithology* 123, no. 2 (June 2011): 373–77.

Chapter 7. An Owl in the Hand

International Association of Avian Trainers and Educators, "Position Statement: Welfare of human-reared vs. parent-reared owls in ambassador animal programs" (March 2018), iaate.org/images/article-pdfs/Position_Statement_-_Welfare_of_Human-reared_vs_Parent-reared_Owls_in_Ambassador_Animal_Programs.pdf.

S. Lee et al., "The function of the alula in avian flight," *Scientific Reports* 5 (May 7, 2015): 9914, doi.org/10.1038/srep09914.

V. Nijman and K. A.-I. Nekaris, "The Harry Potter effect: The rise in trade of owls as pets in Java and Bali, Indonesia," *Global Ecology and Conservation* 11 (July 2017): 84–94, doi.org/10.1016/j.gecco.2017.04.004.

Raptor Center, College of Veterinary Medicine, University of Minnesota, raptor.umn.edu.

Wildlife Center of Virginia, wildlifecenter.org.

Chapter 8. Half Bird, Half Spirit

N. Ashley and M. Sichilongo, eds., *Owls Want Loving: A Reader* (Lusaka, Zambia: Zambian Ornithological Society, 1999).

P. Benavides and J. T. Ibarra, "Uncanny creatures of the dark: Exploring the role of owls across human societies," *Anthropos* 116, no. 1 (2021): 179–92.

M. Bonta, *Seven Names for the Bellbird: Conservation Geography in Honduras* (College Station: Texas A&M University Press, 2003).

I. M. Braun, "Representations of birds in the Eurasian Upper Palaeolithic Ice Age art," *Boletim do Centro Português de Geo-História e Pré-História* 1, no. 2 (November 2018): 13–21.

M. Cocker and H. Mikkola, "Owls and Traditional Culture in Africa," *Tyto* 5, no. 4 (2000): 174–86.

M. Cocker and H. Mikkola, "Magic, myth and misunderstanding: Cultural responses to owls in Africa and their implications for conservation," *Bulletin of the African Bird Club* 8, no. 1 (March 2001): 30–35.

J. A. Forsythe, "Owls & owl feathers in Native American culture," *Whispering Wind* 48, no. 2 (2019): 16–24.

J. Haw et al., "Owl education and conservation in South Africa—successes of 20 years (owlproject.org)," abstract, in "Proceedings of the 6th World Owl Conference, Pune, India, 2019," *Ela Journal of Forestry and Wildlife* 11, no. 1 (January–March 2022): 1209, elafoundation.org/ela/wp-content/uploads/2022/08/EJFW-11-1.pdf.

K. Hull and R. Fergus, "Birds as seers: An ethno-ornithological approach to omens and prognostication among the Ch'Orti'Maya of Guatemala," *Journal of Ethnobiology* 37, no. 4 (December 2017): 604–20.

E. S. Hunn and T. F. Thornton, "Tlingit Birds: An Annotated List with a Statistical Comparative Analysis," in *Ethno-ornithology: Birds, Indigenous Peoples, Culture and Society*, ed. S. C. Tidemann and A. Gosler (London: Routledge, 2010), 181.

S. T. Hussain, "Gazing at owls? Human-strigiform interfaces and their role in the construction of Gravettian lifeworlds in East-Central Europe," in "Archaeo-ornithology: Emerging Perspectives on Past Human-Bird Relations," ed. C. Kost and S. T. Hussain, special issue, *Environmental Archaeology* 24, no. 4 (2019): 359–76.

S. T. Hussain, "The hooting past: Re-evaluating the role of owls in shaping human-place relations throughout the Pleistocene," *Anthropozoologica* 56, no. 3 (February 2021): 39–56, doi.org/10.5252/anthropozoologica2021v56a3.

J. T. Ibarra et al., "Winged voices: Mapuche ornithology from South American temperate forests," *Journal of Ethnobiology* 40, no. 1 (2020): 89–100.

J. Iskandar, B. S. Iskander, and R. Partasasmita, "The local knowledge of the rural people on species, role and hunting of birds: Case study in Karangwangi Village, Cidaun, West Java, Indonesia," *Biodiversitas* 17, no. 2 (October 2016): 435–46.

H. Kettunen, "Owls in Mesoamerica," in "Owls in Myth and Culture," ed. D. H. Johnson (unpublished manuscript, 2020).

V. Laroulandie, "Owls and Hunter-Gatherers in the Upper Paleolithic of France: Evidence from Bone Remains and Art," in "Owls in Myth and Culture," ed. D. H. Johnson (unpublished manuscript, 2020).

K. Macintyre, B. Dobson, and I. Hayward-Jackson, "Owl Voices as Night Spirits: An Ethno-ornithological Approach to the Understanding of the Significance of Night

Bird Calls and Social Control in Traditional Nyungar Culture," in "Owls in Myth and Culture," ed. D. H. Johnson (unpublished manuscript, 2020).

H. Mikkola, "Owl beliefs in Kyrgyzstan and some comparison with Kazakhstan, Mongolia and Turkmenistan," in *Owls*, ed. H. Mikkola (London: IntechOpen, 2019), doi.org/10.5772/intechopen.88711.

S. Molares and Y. Gurovich, "Owls in urban narratives: Implications for conservation and environmental education in NW Patagonia (Argentina)," *Neotropical Biodiversity* 4, no. 1 (2018): 164–72, doi.org/10.1080/23766808.2018.1545379.

M. N. Muiruri and P. Maundu, "Birds, People and Conservation in Kenya," in *Ethno-ornithology: Birds, Indigenous Peoples, Culture and Society*, ed. S. Tidemann and A. Gosler (London: Routledge, 2010), 279–89.

P. Nivedita et al., "Current perceptions of owls held by residents of rural Western Maharashtra, India," in "Proceedings of the 6th World Owl Conference, Pune, India, 2019," *Ela Journal of Forestry and Wildlife* 11, no. 1 (January–March 2022): 1182–84, elafoundation.org/ela/wp-content/uploads/2022/08/EJFW-11-1.pdf.

G. Pam, D. Zeitlyn, and A. Gosler, "Ethno-ornithology of the Mushere of Nigeria: Children's knowledge and perception of birds," *Ethnobiology Letters* 9, no. 2 (2018): 48–64, doi.org/10.14237/ebl.9.2.2018.931.

S. Pande et al., "Owl education and conservation in India: Ela Foundation experience," in "Proceedings of the 6th World Owl Conference, Pune, India, 2019," *Ela Journal of Forestry and Wildlife* 11, no. 1 (January–March 2022): 1221, elafoundation.org/ela/wp-content/uploads/2022/08/EJFW-11-1.pdf.

S. Rajurkar et al., "Owls and cemeteries: Owls are not ghosts," in "Proceedings of the 6th World Owl Conference, Pune, India, 2019," *Ela Journal of Forestry and Wildlife* 11, no. 1 (January–March 2022): 1172–73.

R. Rozzi et al., *Multi-ethnic Bird Guide of the Sub-Antarctic Forests of South America* (Denton: University of North Texas Press and Ediciones Universidad de Magallanes, 2010).

N. Sault, "For the birds, part II: How birds show us the advantages of an ethnobiological perspective," *Journal of Ethnobiology* 37, no. 4 (December 2017): 601–3.

S. Tidemann and A. Gosler, eds., *Ethno-ornithology: Birds, Indigenous Peoples, Culture and Society* (London: Routledge, 2011).

M. Walsh, "Birds of omen and little flying animals with wings," *East Africa Natural History Society Bulletin* 22, no. 1 (March 1992): 2–9.

D. Williams, *Ainu Ethnobiology*, Contributions in Ethnobiology, ed. M. Quinlan and J. Nolan (Tacoma, WA: Society of Ethnobiology, 2017).

F. S. Wyndham and K. E. Park, "'Listen carefully to the voices of the birds': A comparative review of birds as signs," *Journal of Ethnobiology* 38, no. 4 (December 2018): 533–49.

Chapter 9: What an Owl Knows

A. Agarwal et al., "Spatial coding in the hippocampus of flying owls," bioRxiv.org, preprint, submitted October 24, 2021: doi.org/10.1101/2021.10.24.465553.

R. J. Bilney, "Poor historical data drive conservation complacency: The case of mammal decline in south-eastern Australian forests," *Austral Ecology* 39, no. 8 (December 2014): 875–86.

R. J. Bilney, R. Cooke, and J. G. White, "Underestimated and severe: Small mammal decline from the forests of south-eastern Australia since European settlement, as revealed by a top-order predator," *Biological Conservation* 143, no. 1 (January 2010): 52–59.

P. Halme et al., "Do breeding Ural owls *Strix uralensis* protect ground nests of birds?: An experiment using dummy nests," *Wildlife Biology* 10, no. 2 (June 2004): 145–48.

S. Herculano-Houzel, "Birds do have a brain cortex—and think," *Science* 369, no. 6511 (September 25, 2020): 1567–68, doi.org/10.1126/science.abe0536.

M. D. Johnson and D. St. George, "Estimating the number of rodents removed by barn owls nesting in boxes on winegrape vineyards," *Proceedings of the Vertebrate Pest Conference* 29 (August 2020): 1–8.

J. C. Mueller et al., "Evolution of genomic variation in the Burrowing Owl in response to recent colonization of urban areas," *Proceedings of the Royal Society B* 285 (May 16, 2018): 20180206, doi.org/10.1098/rspb.2018.0206.

S. Olkowicz et al., "Birds have primate-like numbers of neurons in the forebrain," *PNAS* 113, no. 26 (June 28, 2016): 7255–60.

M. Stacho et al., "A cortex-like canonical circuit in the avian forebrain," *Science* 369, no. 6511 (September 25, 2020): eabc5534, doi.org/10.1126/science.abc5534.

Afterword: Saving Owls

E. Buechley et al., "Global raptor research and conservation priorities: Tropical raptors fall prey to knowledge gaps," *Diversity and Distributions* 25 (2019): 856–86, doi.org/10.1111/ddi.12901.

J. R. Duncan, "An evaluation of 25 years of volunteer nocturnal owl surveys in Manitoba, Canada," in "World Owl Conference 2017," ed. Inês Roque, special issue, *Airo* 29 (2021): 66–82.

S. T. Garnett et al., "Did hybridization save the Norfolk Island Boobook Owl *Ninox novaeseelandiae undulata*?," *Oryx* 45, no. 4 (October 2011): 500–504, doi.org/10.1017/S0030605311000871.

Government of Nepal, Ministry of Forests and Environment, *Owl Conservation and Action Plan for Nepal 2020–2029* (Kathmandu, Nepal, 2020).

D. F. Hofstadter et al., "Arresting the spread of invasive species in continental systems," *Frontiers in Ecology and the Environment* 20, no. 5 (June 2022): 278–84, doi.org/10.1002/fee.2458.

E. Korpimäki, "Habitat degradation and climate change as drivers of long-term declines of two forest-dwelling owl populations in boreal forest," in "World Owl Conference 2017," ed. Inês Roque, special issue, *Airo* 29 (2021): 278–90.

A. C. Lees et al., "State of the World's Birds," *Annual Review of Environment and Resources* 47 (2022): 231–60.

P. D. Olsen, "Re-establishment of an endangered subspecies: The Norfolk Island Boobook Owl *Ninox novaeseelandiae undulata*," *Bird Conservation International* 6, no. 1 (March 1996): 63–80.

P. D. Olsen et al., "Status and Conservation of the Norfolk Island Boobook *Ninox novaeseelandiae undulata*," in *Raptors in the Modern World*, ed. B.-U. Meyburg and R. D. Chancellor (Berlin: World Working Group on Birds of Prey and Owls, 1989), 123–29.

P. Saurola, "Bad news and good news: Population changes of Finnish owls during 1982–2007," in "Proceedings of the Fourth World Owl Conference, 31 October–4 November 2007, Groningen, the Netherlands," ed. D. H. Johnson, D. Van Nieuwenhuyse, and J. R. Duncan, *Ardea* 97, no. 4 (December 2009): 469–82.

F. Sperring et al., *Ecology, Genetics and Conservation Management of the Norfolk Island Morepork and Green Parrot: Interim Report* (Brisbane: NESP Threatened Species Recovery Hub Project, 2021).

D. J. Tempel et al., "Population decline in California Spotted Owls near their southern range boundary," *Journal of Wildlife Management* 86, no. 2 (February 2022): e22168, doi.org/10.1002/jwmg.22168.

Illustration Credits

Interior Illustrations:

Pages iv and xi: Copyright © Jeff Grotte

Page 4: Courtesy of Nick Athanas / Tropical Birding Tours

Page 8: Courtesy of Lynn Bystrom Photography

Page 14: Courtesy of Ambika Angela Bone

Page 15: Courtesy of Matt Poole

Page 21: Copyright © Roar Solheim / Norsk Naturreportasje

Page 33: Copyright © Melissa Groo

Page 34: Courtesy of Ryan P. Bourbour

Page 36: Image by Gary Gray from Pixabay

Page 41: Copyright © Jeff Grotte

Page 49: NASA

Page 55: Courtesy of Jonathan C. Slaght

Page 60: Copyright © José Carlos Motta-Junior

Page 69: Compliments of David H. Johnson, Global Owl Project

Page 73: Courtesy of Nathalie Morales

Page 80: Courtesy of Lynn Bystrom Photography

Page 97: Copyright © Roar Solheim / Norsk Naturreportasje

Page 102: Copyright © Marjon Savelsberg, reproduced by permission

Page 110: Copyright © Marjon Savelsberg, reproduced by permission

Page 113: Copyright © Daniel J. Cox / naturalexposures.com

Page 114: Courtesy of Julie Kacmarcik

Page 125: Copyright © Jeff Grotte

Page 133: Copyright © Melissa Groo

Page 141: Courtesy of Nick Bradsworth

Page 143: Courtesy of Kurt Lindsay

Page 149: Copyright © Daniel J. Cox / naturalexposures.com

Page 159: Courtesy of Lynn Bystrom Photography

Page 161: Copyright © Melissa Groo

Page 166: Copyright © Jeff Grotte

Page 169: Courtesy of Čeda Vučković

Page 173: Courtesy of Čeda Vučković

Page 179: Copyright © Jeff Grotte

Page 188: Copyright © Jeff Grotte

Page 195: Copyright © Roar Solheim / Norsk Naturreportasje

Page 198: Copyright © Roar Solheim / Norsk Naturreportasje

Page 204: Stipple engraving by F. Holl, 1855, after Parthenope Nightingale, Wellcome Collection

Page 207: Copyright © 2006 by Laura Erickson

Page 221: Courtesy of Wildlife Center of Virginia

Page 225: Courtesy of Wildlife Center of Virginia

Page 233: Gift of Mr. and Mrs. Nathan Cummings, 1964. The Metropolitan Museum of Art

Page 238: Purchase, Lila Acheson Wallace Gift, 1997. The Metropolitan Museum of Art

Page 247: Gift of M. Knoedler & Co., 1918. The Metropolitan Museum of Art

Page 250: Mr. J. H. Wade, gifted to the Cleveland Museum of Art

Page 252: Charles Stewart Smith Collection, Gift of Mrs. Charles Stewart Smith, Charles Stewart Smith Jr., and Howard Caswell Smith, in memory of Charles

Stewart Smith, 1914. The Metropolitan Museum of Art

Page 255: Courtesy of Lynn Bystrom Photography

Page 258: Courtesy of Lynn Bystrom Photography

Page 271: Copyright © David Lei

Page 275: Copyright © Roar Solheim / Norsk Naturreportasje

Page 283: Courtesy of Jonathan Haw, owlproject.org

Page 288: Courtesy of Flossy Sperring

Page 293: Copyright © 2022 Day Scott

Insert Illustrations:

Page 1: Courtesy of Matt Poole

Page 2 (top): Copyright © Roar Solheim / Norsk Naturreportasje

Page 2 (bottom): Courtesy of Matt Poole

Page 3 (top left): Courtesy of Nathan Clark

Page 3 (center right): Copyright © Roar Solheim / Norsk Naturreportasje

Page 3 (center left): Courtesy of Nick Bradsworth

Page 3 (bottom right): Copyright © Roar Solheim / Norsk Naturreportasje

Page 4 (top): Copyright © Roar Solheim / Norsk Naturreportasje

Page 4 (bottom): Copyright © Jeff Grotte

Page 5 (top right): Copyright © José Carlos Motta-Junior

Page 5 (center left): Copyright © Roar Solheim / Norsk Naturreportasje

Page 5 (center right): Courtesy of Čeda Vučković

Page 5 (bottom left): Copyright © Roar Solheim / Norsk Naturreportasje

Page 6 (top left): Gift of Judith H. Siegel, 2014. The Metropolitan Museum of Art

Page 6 (top right): Gift of Citizens' Committee for the Army, Navy, and Air Force, 1962. The Metropolitan Museum of Art

Page 6 (center left): Rogers Fund, 1907. The Metropolitan Museum of Art

Page 6 (bottom left): Rogers Fund, 1941. The Metropolitan Museum of Art

Page 6 (bottom right): The Michael C. Rockefeller Memorial Collection, Bequest of Nelson A. Rockefeller, 1979. The Metropolitan Museum of Art

ILLUSTRATION CREDITS

Page 7: Copyright © Terresa White. Bronze, 36 x 37 x 18 inches. Used with permission.

Page 8 (top left): Copyright © Brad Wilson Photography

Page 8 (top right): Copyright © Brad Wilson Photography

Page 8 (bottom left): Copyright © Brad Wilson Photography

Page 8 (bottom right): Copyright © Brad Wilson Photography

Index

Italicized page numbers indicate material in photographs or illustrations.

Acharya, Raju, 281
Alagoas Screech Owl, 5, *274*
anatomy
  brain, 30–31, 259–60
  ears, 22–23, 30–31
  eyes, 27–30
  facial disks, 20–21, *21*, 44–45
  feathers/plumage, 38, 39, 40
  flight and, 214
  hunting and, 17–18
Arent, Lori, 230–31, 264
Arizona, xvii, 61, 134, 290–91
art, owl portrayals in
  ancient Egyptians and, xi, *238*, 248, 249
  ancient Greeks and, 249–50, *250*
  Chauvet Cave paintings, France, xi, 233–34
  death and, *247*
  human depravity and, 250–51
  Japanese, xi, 251–52, *252*
  the Met, collection of, 248–51
  Moche people and, *233*, 250
  northern European, 250–51
  Snowy Owl, 234–35
Athena (goddess), 203, 249–50
attracting owls. *See* finding and attracting owls
attributes and behaviors. *See also* *specific owls and topics*
  camouflage, 7–8, *8*, 10–11, 18, 35, 47, 48, 130, *255*, 255–56
  distinct from other birds, 2
  head turning, 28
  as human helpers, 266–67
  identification ease and, 1
  individuality of, 93–94, 227
  interpretation of, 223–30
  life span, 230–31, 258
  monogamy, xiv, xvi
  owl attacks on humans, 153–54, 245–46
  playfulness, 262–63
  secretiveness and mysteriousness, xii–xiii, xvi, 224–25
  as teachers, xviii, 226, 256, 267–69
  territoriality, 48, 88–89, 92–93, 94, 95–96, 124
  variety across species, xv–xvi, 22, 161
  variety within species, 74–77
Australia, xvi, 13, 51, 56, 82. *See also* Powerful Owl
  extinctions in, 268–69
  nests in, 140–42
  Norfolk Island Morepork conservation in, 287–89, *288*
Australian Masked Owl, xvi

baby owls. *See* chicks and rearing
Bala, Avinash Singh, 30

INDEX

banding owls, 7, 11
 migration studies and, 181–90, 192–93
 Northern Saw-whet Owl and, xviii, 181–90, 192–93, 194
 Project Owlnet, 181–82, 186–90
 volunteering for, 181–82, 278
Bare-legged Owl, 3
Barking Owl
 capturing, 57
 vocalizations, 82
Barlow, Jon, 132–33, 142–43
barn owls, *283*
 beliefs about, 244–45
 in captivity, 215
 chicks, *114*, 114–15
 courting and breeding, 122
 ear anatomy, *22–23*
 feather pigmentation, 33–35, *34*
 in flight, 36, *36*
 hearing, 30
 hunting by sound, 19–20, 24, 25
 nests, 114–15, 128
 playfulness, 263
 roosts, 175
 species within, 4
 in Virginia, *114*, 114–15, 128
 vocalizations, 82
Barred Owl, 177, 273
 audio recordings, 53
 camouflage, *255*
 in captivity, 215, 223–24, *225*
 Central Park's Barry, 217, 271, *271*
 chicks and rearing, 161, 264
 courtship behavior, 121–22, 124
 dogs used in detection of, 49–50
 ears, 22
 humans attacked by, 245–46
 intelligence, 261–62
 native and expanded habitat, 52
 poisoning threats, 217, 218
 release of rehabilitated, 269–71, *271*
 spotted owls, interactions with, 6, 48, 52–54
 technology in studying, 6
 vocalizations, 53–54, 82, 121, 122, 124
bears, 129–30, 138–39, 142
behavior. *See* attributes and behaviors; *specific owls and attributes*
beliefs and mythology, owl
 in Africa, 236–37, 245
 in ancient Greece, xi, 240, 249–50, *250*
 avimancy, 238–41
 barn owls and, 244–45
 biocultural memory, 242–43
 Blakiston's Fish Owl and, 236
 in Brazil, 269
 death and fear in, xii, 236–37, 239, 244–47, *247*, 253, 284
 diversity and dichotomy of, 252–53
 Ferruginous Pygmy Owl and, 240
 good omens, owls as, 237–39, 243–44
 Great Horned Owl, 240
 human depravity and, 250–51
 in India, xii, 246–47, 281
 Indigenous peoples, xii, 236–37, 240, 242–43, 253, 269
 in Japan, 236–37
 Little Owl, 284
 Mottled Owl, 237, 240
 owl attacks on humans and, 245–46
 population threats, relation to, 281–82, 284
 in Serbia, 171–72
 supernatural and superstitious, xii, 3, 171–72, 236–41, 244–45, 248, 252–53
 survey across cultures of, 243–44, 269
 weather prediction and, 240
Bennett, James Gordon, Sr., 204–5
Bierregaard, Rob, 6, 82, 261–62, 263
Bilney, Rohan, 268
bird counts, 172–73, 279–80
bird markets, 211–13
Blakiston's Fish Owl, *55*
 beliefs and myths about, 236
 capturing, 57–58
 conservation efforts, 286–87
 courtship vocalizations, 121
 hunting by sound, 20
 rearing other owl chicks, 111
 technology for studying, 54–56
Bloem, Karla
 background, 85

owl rehabilitation and education work, 79–94, 208–9, 263
on owl sleep, 160
on population threats, 280
vocalizations studies, 80–81, 83, 84–92, 98, 103
Boobook Owl, 3, 39, 62, 140, 160
Boreal Owl, 20
conservation efforts, 278–79
ears, 22
habitat, xviii, 118
migration, 193–94, 197
brain, 30–31, 258–60
Brazil
Burrowing Owl studies in, 58–59, 62–66, 72–76
owl beliefs in, 269
owl diversity in, xvii, 61–62
screech owls in, 5, 9, 274
Brinker, David, 181, 192
Brown Fish Owl
scavenging, 16
vocalizations, 83
Buff-fronted Owl, 61
Buhl, Gail, 226–31, 260, 262, 264
Burrowing Owl, 3, 7, *60*
capturing, 58–60, 63, 66, 68–69, 71–74, *73*
characteristics, 59, 74
chicks, 154–55
courting and breeding, 122, 126–27
eating and food habits, 65–66
feathers, 9, 73
genetics, 62, 76
geotagging and tracking, 69–71
habitat and range, 60, 62
human-made burrows for, 67–68, 69
hunting time, 59
intelligence, 265
nests, xvi, 59, 65, 128, 131–32
pellet regurgitation, *15*
population threats, 277
variety within species, 74–76

California Spotted Owl
Barred Owl, interaction with, 52–54
vocalizations, 6
calls, owl. *See* vocalizations
cameras, 5–6, 16, 41, 144–45, 155
captivity, owls in
as ambassador birds, 206, 223–25, 226, 230–31
barn owls, 215
Barred Owl, 215, 223–24, *225*
bird markets and, 212–13
challenges of keeping, 206–7, 231
chicks, 219–22, *221*, 286
courting and breeding of, 90–92, 93–94
Eastern Screech Owl, 206–7, *207*, 215, 218, 223, 225
for education purposes, 79–94, 206, 208–9, 211–12, 213–31, 263, 269–71
emotional bonds with humans, 208–9
emotions and psychology, understanding from, 227–30
euthanasia for, 219
famous people and, xvii, 203–6, *204*, 248
feeding, 206–7, 213–14, 227
Great Horned Owl, 79–81, *80*, 84–91, 94, 208–9, 219, *221*
human-imprinted, 79–80, *80*, 84–88, 94, 206–9, 213, 219–20, 231, 245–46
hunting and prey for, 222–23
individual vocalizations, 93–94
Long-eared Owl, 208
Nightingale, Florence, and, 203–4, *204*
Northern Saw-whet Owl, 226–27
not human-imprinted, 87–88, 90–91
regulations and market around, 211–13
rehabilitation and, 79–94, 208–9, 213–29, 263, 269–71
releasing into wild, 269–71, *271*
as surrogate parents, 219–21, *221*
territorial vocalizations, 93
training, 210, 221–30
capture of
Burrowing Owl, 58–60, 63, 66, 68–69, 71–74, *73*
Elf Owl, 292–93
traps and methods for, 56–58

Chauvet Cave paintings, France, xi, 233–34
chicks and rearing. *See also* juveniles; nests
  abandonment or neglect, 155–56
  altruism among siblings, 157–58
  barn owls, *114*, 114–15
  Barred Owl, 161, 264
  branching, 158–59, *159*
  Burrowing Owl, 154–55
  in captivity, 219–22, *221*, 286
  care, 149–52, 155–56
  defensive behavior, 154–55, 160–61
  eating and food habits, 151–52, 155, 157–58, 159–60
  Elf Owl, 151
  Eurasian Eagle Owl, 100–101, 106, 109–11, *110*
  falling out of nests, 137, 156, 157
  feathers, toll of damaged, 152
  first flights and, 159, *161*
  Flammulated Owl, 151
  Great Gray Owl, 150, 152, 153
  Great Horned Owl, 150, 155–56, 161, *161*, 219–21, *221*, 264
  hatching of, 146–47
  hunting, learning, 160–61
  independence, variety across species, 161
  Long-eared Owl, 150
  males provide food for, 151–52
  nest cam insights on, 6, 144, 155
  Northern Pygmy Owl, xvi, *113*, 145–49, *149*
  owls adopting other birds, 110–11
  protection of, 152–55
  sex roles in, 151–53
  siblings eating one another, 156–57
  sleep, xiv, 160
  snakes and, 151
  Snowy Owl, 151, 157
  surrogate parent owls in captivity for, 219–21, *221*
  threats, 130, 135, 140, 147, 150–51, 155, 158, 161, 220, 267
  vocalizations, 89, 91–92, 106
Chile, 240, 242–43, 266
Chocolate Boobooks, 3
citizens. *See also* volunteering
  conservation help from, 172–73, 279–80, 289–92
  research impacts of, xvii–xviii, 98, 119, 279–80, 289–92
Clark, Christopher
  bird flight studies, 37–38
  on feather sound, 38–39, 40
  owl hunting studies, 40–41, 43
Clarke, Rohan, 287–88
Cloud Cuckoo Land (Doerr), 267–68
common names, about, xiv*n*
conservation
  bird counts aiding, 172–73, 279–80
  citizens aiding, 172–73, 279–80, 289–92
  climate change and, 7, 141–42, 192, 266, 274, 277, 290–91
  education's impact on, 281–85, *283*
  efforts and tools for, 278–84
  Elf Owl, 291, 292–93
  festival around, 285
  human activity threats, relation to, 274–75, 277, 280–81
  in India, 246
  Indigenous peoples and, 269
  Norfolk Island Morepork, 287–89, *288*
  rodenticides education and, 283–84
  in Serbia, 172
  South Africa project on, 282–84, *283*
  species at risk and in need of, 273–76
  threats around beliefs and mythology, 281–82, 284
  timing for, 277–78
Cook, Kim, 182, 183–88, 189
courting and breeding
  barn owls, 122
  Barred Owl, 121–22, 124
  Burrowing Owl, 122
  in captivity, 90–92, 93–94
  Eurasian Eagle Owl, xvii, 108–9
  extreme behavior during, 121–22
  feathers, 121, 125–26
  flight behavior during, 124–25
  food and, 121, 122–23
  Great Gray Owl, 120–21, 122–23

Great Horned Owl, 93–94
Long-eared Owl, 123
mate age and size, role in, 126–27
monogamous behavior, xiv, xvi
Northern Pygmy Owl, xviii, 115–16, 117, 120
ORI studies on, 119
Short-eared Owl, 123–25
Snowy Owl, 119, 125–26, 196–97
vocalizations, 86, 90–91, 115–16, 117, 118, 120–21
Czech Republic, 95–96

Difficult Bird Research Group, 51
Doerr, Anthony, 267–68
dogs, for finding owls, 6–7, 49–52
drones, 5, 54–56, 124
Duncan, Jim
on chicks and rearing, 147, 150–51, 153
on conservation efforts, 290
on facial disks, 21, 44–45
Great Gray Owl studies, 35, 40–41, 122–23, 153, 263–64
Long-eared Owl cared for by, 208
on Long-eared Owl parenting, 150
on Northern Saw-whet Owl, 179–80
owl survey work, 279–80

eagle owls. *See* Eurasian Eagle Owl; Verreaux's Eagle Owl
ears. *See also* hearing
actual location of, 22
anatomy, 22–23, 30–31
naming of owls and, 21–22
variety across species, 22
Eastern Screech Owl, 273
camouflage, 8, 47
in captivity, 206–7, *207*, 215, 218, 223, 225
car threats, 216
courtship vocalizations, 121
ears, 24
hunting and prey, 151
lead poisoning of, 217–18
misinterpreting behavior of, 225
vocalizations, 82, 121
eating and food habits. *See also* hunting and prey; pellets
Burrowing Owl, 65–66
cannibalism, 156–57
captivity and, 206–7, 213–14, 227
chicks, 151–52, 155, 157–58, 159–60
courtship, 121, 122–23
Flammulated Owl, 15, 264–65
Great Gray Owl, 16, 18–19
nest cam insight into, 6, 144, 155
scavenging, xiv, 16
stockpiling, 17
vocalizations around chicks, 89–90
way of handling, 13–14
education efforts
Bloem's work in rehabilitation and, 79–94, 208–9, 263
for conservation, 281–85, *283*
Discover Owls, 208
Global Owl Project, xviii, 64, 65, 68
International Owl Center, 80–83, 87–88, 103–4, 263
in Nepal, 281
owls in captivity for, 79–94, 206, 208–9, 211–12, 213–31, 263, 269–71
for people with visual disabilities, 284–85
on rodenticides, 283–84
in Serbia, 172–73
in South Africa, 246, 281–84, *283*
eggs
care, 149–50
color, 145–46
number, 145
shape, 146
threats, 150
timing of laying, 146–47
Egyptians, ancient, xi, *238*, 248, 249
Elf Owl, 3, *293*
capturing, 292–93
chicks and rearing, 151
hunting and prey, xiv, 151
Erickson, Laura, 206–7, 209, 210–11, 219
Esclarski, Priscilla, 64–65, 73–74, 76, 269

INDEX

Eurasian Eagle Owl, *102*
- art portrayal of, 249
- attributes, 3, 100–102
- chicks and parenting, 100–101, 106, 109–11, *110*
- conservation efforts, 278–79
- courting and breeding, xvii, 108–9
- feathers, 39
- flight, 36
- habitat and range, 101
- hunting and prey, 3, 100
- monitoring challenges, 101–2
- nests, 101
- population threats, 101
- roosts, 165
- scavenging, 16
- vocalizations, 98–100, 102–9

evolution
- different habitats affecting, 75
- feathers and, 32–33
- Paleocene appearance and, 1–2
- pellets and, 15
- vision and, 28–29

extinction. *See* conservation

eyes. *See also* vision
- forward-facing, 27–28
- pupils, 29–30

facial disks, 20–21, *21*, 44–45

famous owl keepers
- Bennett, 204–5
- Nightingale, xvii, 203–4, *204*
- Picasso, xvii, 205–6, 248
- Roosevelt, xvii, 203

Fearful Owl, 3

feathers/plumage
- abundance of, 32, 34–35
- age and, 187
- anatomy, 38, 39, 40
- brood patch, 136–37
- Burrowing Owl, 9, 59, 73
- camouflage and, 35
- courting and breeding and, 121, 125–26
- Eurasian Eagle Owl, 39
- evolution and adaptations of, 32–33
- Great Gray Owl, 35, 167
- Great Horned Owl, 34–35, 167
- insulation of, 167–68
- Long-eared Owl, 35, 167
- molting, 34, 152
- parenting's toll on, 152
- pigmentation, 33–35, *34*
- quiet flight, relation to, 32, 35–36, 38–40, 42, 292
- rehabilitation of damaged, 216–17
- Snowy Owl, 35, 125–26, 167

Ferruginous Pygmy Owl
- beliefs about, 240
- hunting and prey, 15–16
- population threats, 274

finding and attracting owls
- challenges in, 47, 255–56
- dogs used for, 6–7, 49–52
- drone technology for, 5, 54–56, 124
- etiquette when, 289–90
- for nest studies, 129–34
- new tools for, 7–9
- sunglasses and, 72
- vocalizations for, 8–9, 48–49, 50, 66, 68–69, 71, 75, 138, 183

Finland, 9, 154, 267, 278–79

Flammulated Owl
- camouflage, 47
- chicks and rearing, 151
- conservation efforts, 277, 291
- eating habits, 15, 264–65
- intelligence, 264–65
- migration, 192
- nests, 128
- vocalizations, 82, 83–84

Fleming, Robyn, 248–49, 251–52

flight
- anatomy and, 214
- audible signature of, 38–39
- barn owl in, 36, *36*
- chicks' first, 159, *161*
- courtship, 124–25
- Great Gray Owl, *32*, 40–41, *41*
- hummingbirds, 37–38
- Northern Saw-whet Owl, 3, *188*
- other raptors compared with owl, 36–37, 38–39
- quiet, 32, 35–36, 38–40, 42, 292

Short-eared Owl, 124–25, *125*
teaching captive owls about, 221–22
food. *See* eating and food habits; hunting and prey; pellets
France
ancient owl remains found in, 234–35
Chauvet Cave paintings of owls in, xi, 233–34

genetics
Burrowing Owl, 62, 76
divergence of species and, 4–5, 76–77
monogamous behavior and, xiv
parental behavior and, 156
research, 76–77
vision and, 28–29
vocalizations and, 87, 92
Gill, Nicole, 51
Gilot, Françoise, 205–6
Girgente, Diane, 182, 185–87
Global Owl Project, xviii, 64, 65, 68
Graham, Robert Rule, 39, 40
Great Gray Owl
appearance, 257, *258*
camouflage, 47
conservation efforts, 278–79
courtship and breeding, 120–21, 122–23
ears, 24–25
eating habits, 16, 18–19
eyes, 27, 28
facial disk and hearing, 20–21, *21*
feathers, 35, 167
flight, 32, 40–41, *41*
habitat, 118, 136–37
hunting by sound, xv, 20, 23, 31
hunting in snow, 22, 40–42
hunting times, xvi
intelligence, 257, 260, 263–64, 265
migration, 197
nests, 129–30, 135–39, 140, 142–44, *143*, 150
parenting, 150, 152, 153
playfulness, 263
scavenging, 16
vocalizations, 120–21, 138–39
Great Horned Owl
beliefs about, 240
broken-wing display, 90
in captivity, 213–14, 215, 216, 219, 220–21, *221*
chicks and rearing, 150, 155–56, 161, *161*, 219–21, *221*, 264
chicks' flight, *161*
courting and breeding, 93–94
ears, 24
feathers, 34–35, 167
habitat, 86
human-imprinted, 79–81, *80*, 84–87, 94, 208–9, 219
hunting and prey, 16, 48, 86, 213–14
lead poisoning threat, 218
nests, 128
personality, 85–86
as superpredator, 16
territorial aggression, 88–89
vocalizations, 80–81, 84–92, 93–94, 103, 130
wet, *33*
Greeks, ancient
Athena's owl and, 203, 249–50, *250*
beliefs about owls, xi, 240, 249–50, *250*

habitat and range. *See also* migration; roosts
Barred Owl, 52
Blakiston's Fish Owl, 54–56
Boreal Owl, xvi, 118
Brazil, xvii, 61–62
Burrowing Owl, 60, 62
changes, 7
climate change and, 7, 141–42, 192, 266, 274, 277, 290–91
Eurasian Eagle Owl, 101
extent and diversity of global, xii, xvii, 3
Great Gray Owl, 118, 136–37
Great Horned Owl, 86
Long-eared Owl, 118, 164
Mottled Owl, 61
for nests, 127–30
satellite transmitters for understanding, 194–95
Short-eared Owl, 15, 16

habitat and range. *(cont.)*
  Snowy Owl, xvi, 118, 126, 194–96, 198–99, 275–76
  species evolution within different, 75
  Spectacled Owl, xvi, 61
  threats and conservation efforts, 5, 273–76, 286–87
  variety across species, xvi
Harry Potter (character and book series), xi, 203, 209–12
Hartman, Jennifer, 9–10, 48–51
Haw, Jonathan, 246, 282–83
head. *See also* ears; eyes
  facial disks, 20–21, *21*, 44–45
  turning, 28
hearing
  with age, 23
  ear location and, 22
  facial disk's role in, 20–21, *21*, 44–45
  hunting by, xv, 19–21, 23, 24, 25, 31, 43–45
  Konishi studies on, 24–25
  mathematical computations and, 26–27
  multidimensional, 24, 25–26
  nocturnal hunting and, 19–20
  pupil reacting, relation to, 30
  research developments on, xv, 43–44
  unique anatomy around, 22–23
  vision, relation to, xv, 25–26, 30–31
Hernandez, Chloe, 124, 129, 138, 144, 159, 164
Hiro, Steve
  background, 119
  Northern Pygmy Owl research, 115–17, 120, 134, 144–48
  ORI work, 117–18, 119–20
history, human, 233, *233*. *See also* beliefs and mythology, owl; Egyptians, ancient; Greeks, ancient
  ancient owl remains found and, 234–35
  Egyptians and owls in, xi, *238*, 248, 249
  Maya and owls in, xii, 237
  relationship with owls and evolution in, 232, 235–43, 269
Holt, Denver
  background, 118–19
  on conservation, 278
  Long-eared Owl research, 10–11, 164–65, 167–68
  on love of owls, 11
  nest studies under, 127–39, 143–44, 151
  ORI founding and leadership, 117–20
  on roosting behavior, 175, 176
  on sleep behavior, 166
  on Snowy Owl chicks and rearing, 151, 157
  Snowy Owl conservation efforts, 276–77
  on Snowy Owl courting and breeding, 125–26
  on Snowy Owl migration, 196–99, 201
  on Snowy Owl nest defense, 153–54
hoots. *See* vocalizations
human-imprinted owls. *See under* captivity, owls in
hummingbirds
  dive-bombing owls, 10
  flight, 37–38
  as prey, 15–16
hunting and prey
  abilities with, xiv–xv, 13–18, 257
  anatomy and, 17–18
  binocular vision and, 27–28
  Burrowing Owl, 59
  captive owls and, 222–23
  car threats, 216
  chicks learning about, 160–61
  Eastern Screech Owl, 151
  Eurasian Eagle Owl, 3, 100
  fish eaters and, 20, 58
  Great Gray Owl, xv, xvi, 20, 22, 23, 31, 40–42
  Great Horned Owl, 16, 48, 86, 213–14
  Long-eared Owl, 9, 22
  mathematical computations and, xv, 26–27
  migration, relation to, 200–201
  nocturnal, 19–20, 22, 24, 29, 59
  Northern Hawk-Owl, xvi, 16
  Northern Pygmy Owl, xvi, 116, 148
  odds of success with, 16–17

other birds and, xiv, 15–17, 116
Powerful Owl, xvi, 13–14, *14*
quiet flight and, 32, 35–36
rodent control and, 282, 283–84
roosts, relation to, 171, 175–76
Short-eared Owl, 15, 16
in snow, 22, 40–42
Snowy Owl, 16, 151, 196, 197–98, *198*, 200–201, 267
sound/hearing's role in, xv, 19–21, 23, 24, 25, 31, 43–45
technique, 17–18
threats and conservation efforts, 276–77
times, variety across species, xvi
variety across species, 15–16
vocalizations for, xiv

Ibarra, Tomás, 240, 241–43
India
conservation efforts in, 246
Forest Owlet of, 274
Mottled Wood Owl of, 81
owl beliefs and mythology in, xii, 246–47, 281
Indigenous peoples
art portrayals of owls, 250
conservation and, 269
owl beliefs and mythology of, xii, 236–37, 240, 242–43, 253, 269
intelligence
Barred Owl, 261–62
brain anatomy, 259–60
brain size, 258–59
Burrowing Owl, 265
Flammulated Owl, 264–65
flexibility as sign of, 265–66
Great Gray Owl, 257, 260, 263–64, 265
Little Owl, 261
Long-eared Owl, 265–66
mental map ability and, 256–57
migration and, 266
playfulness as sign of, 262–63
tests and measures for, 260–61
International Owl Center, Minnesota, 80–83, 87–88, 103–4, 263
Italy, 16, 81, 246, 284–85

Japan
Blakiston's Fish Owl conservation in, 286–87
owl art from, xi, 251–52, *252*
owl beliefs and mythology in, 236–37
owl import market in, 213
Johnson, David
Burrowing Owl studies, 58–59, 62–77, 126–27, 131–32
on conservation, 277–78, 282
feather counting, 34–35
Global Owl Project, xviii, 64, 65, 68
human-made burrows made by, 67–68
on love of owls, xviii, 63–64
on migration, reasons for, 178, 194
on owls in myths and culture, 235–36, 243–44, 269
on Snowy Owl breeding sites, 197
on territorial aggression, 88–89
on vocalizations, 87, 88–89
juveniles, 92, 109–10, 157, 175, 177, 194

Kacmarcik, Julie, 182–88
Konishi, Masakazu (Mark), 24–25
Köppl, Christine, 22–23, 26

lead poisoning, 217–18
Le Fay, Solai, 136–38, 144
life span, 230–31, 258
Life with Picasso (Gilot), 205–6
Lindo, David, 7, 8–9, 170, 171, 172, 174
Linhart, Pavel, 95–98, 109, 261
Linkhart, Brian, 82, 83–84
literature, owls' appearance in, xii, 277
Cloud Cuckoo Land, 267–68
Harry Potter, xi, 203, 209–12
The Sword in the Stone, 203
Little Owl, *97*, 177
Athena's, 203, 249–50, *250*
beliefs about, 284
individuality of vocalizations, 96–98
intelligence, 261
Nightingale's, 203–4, *204*
Picasso's, 205–6
territorial vocalizations, 92–93, 96

Long-eared Owl, *xi*, 7
appearance variability, 173–74
in captivity, 208
courting and breeding, 123
ears, 22
feathers, 35, 167
habitat, 118, 164
hunting, 9, 22
intelligence, 265–66
migration, 163
nests, xii, xiv, 128
parenting, 150
playfulness, 263
roosting in groups, xiv, xvi, 164, 167–76, *169*, *173*
scavenging, 16
territorial behavior, 124
in urban environments, 170–71, 175–77, 265–66
vocalizations, 83, 84, 174, 208
Long-tufted Screech Owl, 61–62
Long-whiskered Owlet, 3, *4*, 274

Magdalenian culture, 234
Martin, Graham, 27, 30
Mastrorilli, Marco, 246, 284–85
Maya, xii, 237
Mendelsohn, Beth
on chicks and rearing, 156, 158, 161
on chicks branching, 158
on nest predation, 155
nest research, 129–30, 135–36, 138–39, 142–43, 144, 155
Mendes, Gabriela, 64–65, 73–74
Merlyn (magician), 203
Metropolitan Museum of Art (the Met), 248–51
Mexico, xvii, 61, 192, 237, 240
migration
banding owls for understanding, 181–90, 192–93
Boreal Owl, 193–94, 197
for breeding sites, 196–98
factors and decisions around, 163–64, 177–78, 193–94, 200–201
Flammulated Owl, 192
hunting and prey, relation to, 200–201
intelligence and, 266
Long-eared Owl, 163
Northern Hawk-Owl, 197
Northern Saw-whet Owl, 163, 178–79, 187–89, 191–93
pace of, 192–93
satellite transmitters for understanding, 194–95
sex and, 193–94
Short-eared Owl, 197
Snowy Owl, 163, 195–201, 266
unpredictability of, xiii–xiv
Minnesota
International Owl Center, 80–83, 87–88, 103–4, 263
Raptor Center, 226–31, 260, 262, 264
Moche people, *233*, 250
molting, 34, 152
monogamous behavior, xiv, xvi
Montana. *See also* Owl Research Institute
species diversity and studies in, 119
species number in, 118
Morepork. *See* Norfolk Island Morepork
mosquitoes, 150–51
Motaung, Kefiloe, 282
Mott, Beth, 140–41, 142, 160–61
Mottled Owl
beliefs about, 237, 240
habitat, 61
meetings at night, xvi
Mottled Wood Owl, 81
mythology. *See* beliefs and mythology, owl

naming, of owls
common, xiv*n*
ears and, 21–22
taxonomy challenges and, 62–63
Navajo myth, xii
Nepal, 281
nests. *See also* eggs
boxes, 122, 135, 140–41, 156, 261, 278–79, 282, 285, 286, 289
broken-wing display for protection of, 90

Burrowing Owl, xvi, 59, 65, 128, 131–32
cameras in, 5–6, 144–45, 155
chicks falling out of, 137, 156, 157
choosing sites for, 127–30, 139–40
co-opting other animals' structures for, 128, 139
decoration of, xiv, 131–32
drones for checking, 56
Eurasian Eagle Owl, 101
female's brood patch and, 136–37
food cache in, 17
ground, xvi, 59, 65, 127–28, 131–32
post-fledgling, 160
protection of, 152–54
research methods and challenges, 128–45
in snags and hollows of trees, 137–38, 140–43, *143*
sex roles in, 71, 128, 131–32
Snowy Owl, 127–28, 150, 151, 153–54, 276
threats, 139–42
Netherlands, 98–99
Nicholson, Amanda, 214, 215, 218, 220, 222–25, 230
Nightingale, Florence, xvii, 203–4, *204*
Norfolk Island Morepork, 32, 287–89, *288*
Northern Hawk-Owl
hunting habits, xvi, 16
migration, 197
Northern Pygmy Owl
attributes, 116–17
chicks, xvi, *113*, 145–48, *149*
courting and breeding, xviii, 115–16, 117, 120
hunting and prey, xvi, 116, 148
nests, 128, 132–34, *133*, 139, 144–49, *149*
vocalizations, 82, 115–16, 117
Northern Saw-whet Owl
aging, 187
appearance and character, 179–81, 185, 186–87, 226
banding of, xviii, 181–90, 192–93, 194
in captivity, 226–27
courtship vocalizations, 121
ears, 24
flight, 3, *188*
hunting by sound, 25
migration, 163, 178–79, 187–89, 191–93
nests, 128, 133–34
research challenges, 189–91, 194
roosts, *179*, 190, 191
secretive nature of, 179, 189–90
sex differences, 186
sleep, *166*
Northern Spotted Owl, *49*, 177
Barred Owl, displacement of, 52–53
camouflage, 48
dogs used in detection of, 49–50
Hartman, research of, 9–10, 48–50
vocalization for attracting, 48–49

offspring. *See* chicks and rearing; juveniles; nests
Oleyar, Dave, 6, 57, 134–35, 141–42, 151, 182, 290–91
ORI. *See* Owl Research Institute
Otus bikegila, 3–4
owl attacks on humans, 153–54, 245–46
owl detection. *See* finding and attracting owls
owlets (baby owls). *See* chicks and rearing
Owl Research Institute (ORI), 124
courting and breeding studies, 119
Great Gray Owl surveys, 129–30
Holt, founding and leadership of, 117–20
Long-eared Owl roosting, 164, 168
nest studies, 129–39, 142–43
Northern Pygmy Owl studies, 115–16
Short-eared Owl studies, 123–24
volunteers, 117, 119–20, 132, 181–82
Owls of the Eastern Ice (Slaght), 54
Owls Want Loving (Zambian Ornithological Society), 282

parenting. *See* chicks and rearing
Park, Karen, 238–39
Payne, Roger, 19–20, 24, 25

pellets
dogs finding, 7, 49–52
other raptors compared with owl, 51
regurgitation, 14–15, *15*
Sooty Owl, 268–69
Stygian Owl, xiv
Peña, José Luis, 26
Picasso, Pablo, xi, xvii, 205–6, 248
Pierce, Karra, 215–18, 220, 221–22, 270–71
playfulness, 262–63
Pleistocene, 2, 234–35
plumage. *See* feathers/plumage
poisoning, of owls
lead, 217–18
rodenticide, 217, 283–84
possum, 13–14, *14*, 16–17
Powerful Owl, 3
capturing, 57–58
fledglings' defensive behavior, 160–61
hunting and prey, xvi, 13–14, *14*
nests, 140–41, *141*, 142
roosts, 165
sibling chicks eating one another, 157
urban sighting of, xvi, 14
predators, of owls, 182
chicks and, 135, 147, 155, 158, 161, 220, 267
defense against, 27–28, 90, 128, 130, 134, 148–49, 155, 164, 165, 175
humans' appearance and, 228, 229
mimicking, in searching for nests. 133–34
Northern Pygmy Owl defense against, 148–49
Short-eared Owl risk from, 155
training captive owls about, 226
prey. *See* eating and food habits; hunting and prey
Project Owlnet, 181–82, 186–90
protection. *See* captivity, owls in; conservation
pygmy owls. *See also* Ferruginous Pygmy Owl; Northern Pygmy Owl
ears, 22
food cache, 17
ranges. *See* habitat and range
Raptor Center, University of Minnesota, 226–31, 260, 262, 264
Rattenborg, Niels, 165–66
rearing. *See* chicks and rearing
research. *See also* captivity, owls in; technology, research; *specific owls and researchers*
archaeozoology, 234–35
audio recording in, 6
banding owls for, xviii, 7, 11, 181–90, 192–93, 194, 278
bird counts aiding, 172–73, 279–80
challenges with, 9–10, 43–44, 47, 54–55, 106–7, 120, 129–31, 135, 144–45, 189–91, 194, 278
citizens' role and impact, xvii–xviii, 98, 119, 279–80, 289–92
conservation-based, 278–79
courting and breeding, 119
dedication to and benefits of, 183–84
dogs used in, 6–7, 49–52
evolution and development, xiii, 1–5
flight, 37–38
genetics, 76–77
hearing, xv, 24–25, 43–44
hunting, 40–41, 43
intelligence tests and measures in, 260–61
interpreting owl behavior, 223–30
migration, 178, 181–93, 200
nests, 128–45, 155
new tools for locating owls for, 7–9
taxonomy and, 62–63
volunteers, 117, 119–20, 132, 181–82
Robb, Magnus, 83
rodenticide poisoning, 217, 283–84
Rogue Detection Teams, 50–51
Roosevelt, Teddy, xvii, 203
roosts
factors and decisions around, 163, 165–68
hunting and prey, relation to, 171, 175–76
as learning centers, 175–77
Long-eared Owl, xiv, xvi, 164, 167–76, *169*, *173*

Northern Saw-whet Owl, *179*, 190, 191
number of owls in, 167
separate species sharing, 175
sleep behavior and, 165–66
Snowy Owl, 175
Rowling, J. K., 203, 209–12
Rufous-legged Owl, 242, 266–67
Ružić, Milan, 169–74, 175, 176–77, 265–66, 284

Sanders, Bernie, 238–39
satellite tracking, 5, 54–56, 70, 194–97, *195*, 199–201, 275
Saurola, Pertti, 278–79
Savelsberg, Marjon
background, 103–4, 105
Eurasian Eagle Owl studies, 98–102, 104–11
saw-whet owls. *See* Northern Saw-whet Owl
Scott, Day, 290–93
screech owls. *See also* Eastern Screech Owl
Alagoas, 5, 274
in Brazil, 5, 9, 274
courting and breeding, 126
Long-tufted, 61–62
nests, 128
population threats, 5, 274, 291
roosts, 165
sibling chicks fighting over food, 156–57
Vermiculated, 3
Whiskered, 291
Xingu, 5, 274
Serbia, *169*, 169–73, *173*, 265–66
Short-eared Owl
appearance and attributes, 228–29
courting and breeding, 123–25
flight, 124–25, *125*
habitat and range, 15, 16
hunting and prey, 15, 16
migration, 197
nest defense, 155
nests, 127, 130, 155
roosts, 175
vocalizations, 123–24
Siau Scops Owl, 274
sight. *See* eyes; vision
Slaght, Jonathan, 20, 54, 57–58, 286–87
sleep
of chicks, xiv, 160
Northern Saw-whet Owl, *166*
unihemispheric, 165–66
snakes, 151
Snowy Owl, 3, *275*
appearance and character, 194, 195–96
archaeozoology and, 234–35
art portrayals of, 234–35
chicks and rearing, 151, 157
courting and breeding, 119, 125–26, 196–97
feathers and plumage, 35, 125–26, 167
habitat and range, xvi, 118, 126, 194–96, 198–99, 275–76
Harry Potter's, xi, 203, 209–12
hunting and prey, 16, 151, 196, 197–98, *198*, 200–201, 267
migration, 163, 195–201, 266
nest defense, 153–54
nests, 127–28, 150, 151, 153–54, 276
as pets, 211, 212
roosts, 175
satellite transmitter tagging of, 194–97, *195*, 201
scavenging, 16
strength of, 210–11
threats and conservation efforts, 275–77
weight, 194
songbirds, hunting of, xiv, 17, 116
Sooty Owl
capturing, 57–58
pellet study, 268–69
roosts, 168
vocalizations, 81
South Africa, 246, 282–84, *283*
species
attribute variety across, xv–xvi, 15–16, 22, 161
divergence of, 4–5, 62–63, 74–76
evolution in different habitats, 75

species *(cont.)*
extinct, 2, 268–69
new and rare, 3–4
number of modern, 3
roosts shared by different, 175
vocalization variety within, 74–77
Spectacled Owl
habitat, xvi, 61
roosts, 165
Sperring, Flossy, 288–89
spotted owls. *See also* California Spotted Owl; Northern Spotted Owl
Barred Owl, interactions with, 6, 48, 52–54
vocalizations, 6, 48–49
Striped Owl, 9, 61
Stygian Owl, xiv, 61
superstitions. *See* beliefs and mythology, owl
Surmach, Rada, 54–56
Surmach, Sergei, 55, 286–87
The Sword in the Stone (White), 203

Tasmanian Masked Owl, 51, 274
Tawny-browed Owl, 3, 61
Tawny Owl, 7
conservation efforts, 284–85
populations, 278–79
vocalizations, 81–82, 95
taxonomy
challenges, 62–63
common names and, xiv*n*
technology, research
acoustic camera and snow hunting, 41
camera traps and nest cams, 5–6, 16, 144–45, 155
drone, 5, 54–56, 124
geotagging, 69–71
for hearing study, 43–44
importance of, 7–8
for nest studies, 129, 144
satellite, 5, 54–56, 70, 194–95, *195*, 199–201, 275
for vocalizations, 36–37, 53, 95, 104, 106–7, 109
Tengmalm's Owl. *See* Boreal Owl
territorial behavior
Great Horned Owl, 88–89
Long-eared Owl, 124
vocalizations and, 48, 88–89, 92–93, 94, 95–96
Therrien, Jean-François, 199–200
training owls, 210, 221–30

Ural Owl, 154, 267, 273, 278–79
urban environments, 160–61, 262, 273
Burrowing Owl in, 60, 65, 265, 266
Long-eared Owl in, 170–71, 175–77, 265–66
Powerful Owl in, xvi, 14

Vermiculated Screech Owl, 3
Verreaux's Eagle Owl, 3
Virginia, 113
barn owls in, *114*, 114–15, 128
Great Horned Owl, first sighting in, 86
Wildlife Center of, 213–24, 269–70
vision
color spectrum and, xv, 29
forward-facing eyes and, 27–28
genetics and adaptation, 28–29
head-turning ability and, 28
hearing, relation to, xv, 25–26, 30–31
vocalizations
algorithms for classifying, 95, 109
Barking Owl, 82
barn owls, 82
Barred Owl, 53–54, 82, 121, 122, 124
Brown Fish Owl, 83
Burrowing Owl, 74–75
California Spotted Owl, 6
chicks and rearing, 89, 91–92, 106
complexity of, 10
courting and breeding, 86, 90–91, 115–16, 117, 118, 120–21
diversity of, 74–77, 81–82
eating and food habits, 89–90
Eurasian Eagle Owl, 98–100, 102–9
feeding chicks and, 89–90
for finding owls, 8–9, 48–49, 50, 66, 68–69, 71, 75, 138, 183
genetics and, 87, 92
Great Gray Owl, 120–21, 138–39

Great Horned Owl, 80–81, 84–92, 93–94, 103, 130
of human-imprinted owls, 87
individuality, 93–98, 108–9
Little Owl, 92–93, 96–98
Long-eared Owl, 83, 84, 174, 208
mimicking, 82–84, 88–89
monitoring of, 52–54, 107–8
Northern Pygmy Owl, 82, 115–16, 117
Northern Saw-whet Owl, 121, 179
Northern Spotted Owl, 48–49
reasons behind, 84–90, 94
sex and, 89, 90–91, 98, 105
Short-eared Owl, 123–24
Sooty Owl, 81
spectrograms of, 36–37, 53, 104, 107
spotted owls, 6, 48–49
Tawny Owl, 81–82, 95
technology for researching, 36–37, 53, 95, 104, 106–7, 109
territorial behavior and, 48, 88–89, 92–93, 94, 95–96
threats and, 90, 154–55
for tricking predators, xiv
variety within species, 74–77
volunteers in owl research, 34
in Arizona, 291
for banding, 181–82, 278
bird counting, 279–80
dedication with, 279–80
emotional/personal benefits of, 119–20
in Finland, 278
impacts of, xviii, 98, 119, 279, 290
New Zealand Moreporks and, 289
at ORI, 117, 119–20, 132, 181–82
in Serbia, 169–70
for vocalization research, 98

Weidensaul, Scott
migration studies, 178, 188–89, 191–92, 193, 200
on Northern Saw-whet Owl research challenges, 189–91, 194
on Snowy Owl hunting, 16
on Snowy Owl migration, 200
on watching owls, 137
Whiskered Screech Owl, 291
White, T. H., 203
White-chinned Owl, 3
Wildlife Center of Virginia, 213–24, 269–70
Wood, Connor, 52–54
Wyndham, Felice, 237–41, 245, 252–53, 269

Xenoglaux, 3–4
Xingu Screech Owl, 5, 274

Yamamoto, Sumio, 111, 286

Zambia, 237, 245, 281–82
Zambian Ornithological Society, 281–82

ALSO AVAILABLE

THE BIRD WAY

A New Look at How Birds Talk, Work, Play, Parent, and Think

"There is the mammal way and there is the bird way." But the bird way is much more than a unique pattern of brain wiring, and what scientists are finding is upending the traditional view of how birds conduct their lives. Drawing on personal observations, the latest science, and her bird-related travel around the world, Jennifer Ackerman shows there is clearly no single bird way of being. In every respect, in plumage, form, song, flight, lifestyle, niche, and behavior, birds vary. It is what we love about them.

THE GENIUS OF BIRDS

As acclaimed author Jennifer Ackerman travels around the world to the most cutting-edge frontiers of research, she not only tells the story of the recently uncovered genius of birds but also delves deeply into the latest findings about the bird brain itself that are shifting our view of what it means to be intelligent. At once personal yet scientific, richly informative, and beautifully written, *The Genius of Birds* celebrates the triumphs of these surprising and fiercely intelligent creatures.

BIRDS BY THE SHORE

Observing the Natural Life of the Atlantic Coast

Birds by the Shore is a book about discovering the natural life at the ocean's edge. Against the landscape of the small coastal town of Lewes, Delaware, Ackerman revisits her own history—her mother's death, her father's illness, and her hopes to have children of her own. With a quiet passion and friendly, generous intelligence, it explores the way that landscape shapes our thoughts and perceptions and shows that home ground is often where we feel the deepest response to the planet.